Soil Fertility Degradation, and Soil Health Restoration and Management

Ram C. Dalal • Somasundaram Jayaraman

Soil Fertility Degradation, and Soil Health Restoration and Management

Springer

Ram C. Dalal
School of Agriculture and Food Sustainability
The University of Queensland
Brisbane, Australia

Somasundaram Jayaraman
ICAR-Indian Institute of Soil and Water Conservation Research Centre
Fern Hill Post
The Nilgiris, Tamil Nadu, India

ISBN 978-981-96-0404-3 ISBN 978-981-96-0405-0 (eBook)
https://doi.org/10.1007/978-981-96-0405-0

This Springer imprint is published by the registered company Springer Nature Singapore Pte Ltd.
The registered company address is: 152 Beach Road, #21-01/04 Gateway East, Singapore 189721, Singapore

Dedicated to
Nirmala Dalal
Vetripriya Somasundaram

Foreword

Soil Fertility Degradation and Soil Health Restoration and Management....

Global population is projected to reach ten billion by the end of 2060. Feeding the growing population and fulfilling food and nutritional security is a daunting task amidst large-scale soil degradation. Moreover, agriculture and food production systems are highly vulnerable to climate change, soil and water scarcity, soil degradation (physical-chemical, and biological degradation) and increasing population pressure, as well as increasing rate of converting fertile lands to non-agricultural purposes. In the quest of increasing food production, animal and human health, and economic growth, humanity has jeopardised/degraded natural resources such as soil, water, air, vegetation, and biodiversity at an alarming rate. Attaining food and nutritional security is not possible without having a good natural resource base, particularly good fertile and healthy soil for sustaining productivity, and delivering ecosystem services while mitigating and adapting to anthropogenic climate change.

Nutrient depletion from soil is a natural process of cultivation of initially fertile soil for farming to provide animal and human food. If the nutrients removed by the crop or in the produce are not replenished, then it leads to *soil fertility degradation*. Besides, the loss in economic produce, large losses of available soil nutrients/loss of soil organic matter, and faunal and microbial biodiversity can occur in a number of ways, such as through soil erosion and runoff, crop residue removal, burning, leaching, volatilisation and denitrification of N, reduced nutrient availability, increased fixation of less mobile elements (inorganic P, Cu, Zn, and Mn), reduced nutrient use efficiency, and microbial diversity due to adverse soil conditions for plant growth (structural degradation, erosion/runoff, acidification, salinisation, sodication, soil-borne diseases, etc.). Consequently, grain yields decline and grain quality deteriorates with increasing periods of cultivation using inappropriate practices if fertility and soil health restorative practices are not used/adopted.

Sustaining soil health is of paramount significance in the present-day scenario as it is directly linked with the 'One Health Concept' (i.e. *Soil-Plant-Animal-Human-Ecosystem-Planetary-Health*). This book is aimed to address all those issues of soil fertility degradation and restoration of soil health via sustainable management practices.

I strongly believe this book will be highly useful to students, scientists, researchers, land managers, and policy makers interested in soil fertility and human nutrition.

I congratulate the authors for bringing out a very good book on *Soil Fertility Degradation, and Soil Health Restoration and Management*.

Sincerely,
Rattan Lal
Distinguished University Professor of Soil Science
Director, CFAES Rattan Lal Center for Carbon Management and Sequestration (Lal Carbon Center), The Ohio State University, Columbus, OH 43210, USA
IICA Chair in Soil Science & Goodwill Ambassador for Sustainable Development Issues
Adjunct Professor at the University of Iceland and the Indian Agricultural Research Institute

Preface

Soils are closely interlinked with human civilisation, wherever good fertile soils exist in the adjoining river system, there exists and spread of larger human population. Therefore, good fertile soil is vital for farming and sustainable production as well as achieving food and nutritional security.

Land degradation (via physical, chemical, and biological processes) is one of the serious threats to global farmland, adjoining landscapes, and riverine systems as well as to ecosystem services. About 33% of global soils are already degraded and this figure may reach much higher by 2050. There is plenty of evidence that soil degradation is still occurring, which further exacerbates the degradation of natural resources, particularly 'Soil'.

Soil fertility degradation occurs mainly due to soil erosion (water and wind), overgrazing, continuous cultivation of farmlands using inappropriate practices, without adding sufficient organics (manures), soil organic matter loss from arable lands, nutrient removal through crop grain or produce/crop residue removal (without proper replenishment)/crop residue burning, acidification, salinisation, sodification, flora, fauna and microbial biodiversity loss, and excess accumulation of heavy metal load and pesticides. All these factors either alone or in combination contribute to higher nutrient depletion (soil organic matter, nitrogen, phosphorus, potassium, sulphur, secondary and micronutrients, etc.) and physical and biodiversity loss, which eventually leads to *soil fertility degradation*. This poses a serious threat to food production, sustainability, and achieving UN's Sustainable Development Goals, especially in tropical and sub-tropical regions.

Thus, it is imperative to identify and monitor one or more processes involved in and affecting the nature of fertility degradation of a given soil so as to devise, promote, and implement sustainable management strategies that would maintain, restore, and, in the long term, enhance the fertility and health of such a degraded soil.

In fact, all the management may not bring proper restoration of soil health, but management practices improve soil organic matter (SOM) or organic carbon and nutrients may bring fertility restoration and enhancement in productivity as the SOM acts as a central role in *soil-fertility-health-productivity-ecosystem services*.

This book addresses all these issues of soil fertility degradation, restoration, and management practices for sustainable production and soil health.

The authors thoughtfully reminisce and express their gratitude to all the teachers and professors who have inspired with their great knowledge and wisdom that has helped and reformed our understanding of Soil Science vis-à-vis Natural Resource Management and Sustainable Soil Health Management. As one teacher appropriately said, 'Soil is the mother of all resources'.

We hope this book will be highly useful to students, researchers, scientists, land managers, and policy makers to devise, advocate, adopt, and practise appropriate management practices/policy decisions and action plans for restoring health of degraded soils particularly fertility depleted soils.

Brisbane, Australia — Ram C. Dalal
The Nilgiris, Tamil Nadu, India — Somasundaram Jayaraman

Acknowledgements

Authors sincerely thank the following persons/organisations

It is always difficult to name each and every person. However, it is our duty to express and place our sincerest thanks to all and everyone, who have directly or indirectly helped us in completing this task.

- Research scholars, PhD students, field assistants, lab-technicians,
- We fondly remember our teachers and professors in infusing Soil Science (specifically but not exclusively, Dr D R Bhumbla, Dr J S Kanwar, Dr N P Datta, Dr R K Chatterjee, ...)
- Staff at Hermitage Research Station, Toowoomba, QLD, Australia
- Staff at the Queensland Wheat Research Institute, Toowoomba (especially Dr Allan Clarke, Dr David Freebairn, Dr Rob Loch, Dr Peter White, Dr John Thompson, Dr Wayne Strong, Errol Weston, Dr Greg Thomas, and Dr Roger Shaw for lively discussions and data sharing)
- School of Agriculture and Food Sustainability, The University of Queensland, Brisbane, Australia
- DSITI, Ecosciences Precinct, Brisbane, Australia
- Indian Council of Agricultural Research [ICAR-Indian Institute of Soil Science (IISS), Bhopal, India and ICAR-Indian Institute of Soil and Water Conservation (IISWC), Research Centre, The Nilgiris, Tamil Nadu, India]
- Family members for giving enough time to complete this assignment.
- We sincerely thank Professor Rattan Lal for writing the Foreword for this book.
- We are thankful to the almighty for giving us good health to complete this work.

Contents

1 Introduction 1

2 Scope of the Book 5

3 Processes of Fertility Degradation 7
- 3.1 Soil Erosion 7
  - 3.1.1 Estimation of Soil Loss 8
  - 3.1.2 Magnitude of Soil Loss 16
  - 3.1.3 Magnitude of Soil Fertility Degradation 18
  - 3.1.4 Soil Erosion and Crop Productivity 23
- 3.2 Nutrient Removal in Produce 27
  - 3.2.1 Natural Ecosystem and Land Clearing 27
  - 3.2.2 Agroecosystem and Nutrient Depletion 27
- 3.3 Nutrient Removal in Crop Residues 30
- 3.4 Oxidation, Reduction, and Leaching 31
  - 3.4.1 Oxidation 31
  - 3.4.2 Reduction 34
  - 3.4.3 Leaching 36
- 3.5 Acidification, Sodification, and Excessive Element Load 39
  - 3.5.1 Acidification 39
  - 3.5.2 Sodification 43
  - 3.5.3 Excess Element Load 51
- 3.6 Aggregate Breakdown 55
  - 3.6.1 Aggregate Formation and Stabilisation in Soil 55
  - 3.6.2 Functions of Soil Aggregates 58
  - 3.6.3 Consequences of Soil Aggregate Breakdown 59
  - 3.6.4 Magnitude of Aggregate Breakdown in Cultivated Soils 61

4 Nature of Fertility Degradation 65
- 4.1 Organic Matter (SOM) Loss 65
  - 4.1.1 Description of Rate of SOM Loss 66
  - 4.1.2 Turnover of SOM in Natural Ecosystems 67
  - 4.1.3 SOM Loss from Cultivated Soils 69
  - 4.1.4 Modelling SOM Turnover 81

4.1.5 Magnitude of SOM Loss from Cultivated Soils 84
4.1.6 Consequences of SOM Loss from Cultivated Soils 85
4.2 Nutrient Depletion 89
4.2.1 Nutrient Eelement Loss 90
4.3 Decline in Nutrient Bioavailability 104
4.3.1 Soil Enzymes 105
4.3.2 Adenosine 5′-Triphosphate and Adenylate Energy Charge 108
4.3.3 Soil Microbial Biomass 109
4.3.4 Mineralisable Nutrients 111
4.4 Soil Physical Degradation 113
4.4.1 Soil Hydrologic Characteristics 113
4.4.2 Soil Aeration 114
4.4.3 Soil Temperature 115
4.4.4 Soil Mechanical Characteristics 116
4.4.5 Interrelationship Among Various Measures of Soil Physical Degradation 118
4.5 Soil Biological Degradation 119
4.5.1 Decrease in Beneficial Soil Microflora and Soil fauna 120
4.5.2 Increase in Harmful Soil Microflora and Soil Fauna 124

5 Fertility-Degraded Soil: Soil Health Restoration and Management-Strategies and Approaches 127
5.1 Soil Physical Restoration 129
5.2 Soil Chemical Restoration 131
5.3 Soil Biological Restoration 134
5.4 Soil Ecological Restoration 135
5.4.1 Principles of Soil Organic Matter Restoration 136
5.4.2 Management Practices for Potential Increase in Soil Organic Matter 139

6 Conclusions 151
6.1 Soil Stewardship Through Sustainable Soil Management and Restoring Soil Health for Fertility of Degraded Soils 151

References 155

About the Authors

Professor Ram C. Dalal AM graduated from the Punjab Agricultural University (1964) and completed his master's and PhD from the Indian Agricultural Research Institute (1969). Presently he is a Professor/Adjunct Professor in the School of Agriculture and Food Sustainability at the University of Queensland, Brisbane, Australia, and an Adjunct Professor at Uttar Banga Krishi Viswavidyalaya, Pundibari, Cooch Behar, India. During the last five decades of research and teaching experience, he has contributed significantly to create awareness in the farming, scientific, and general community to the seriousness and insidious nature of soil/land degradation. For the last 25 years, landscape restoration, soil organic matter management, carbon sequestration and nitrogen management, site-specific management for soil and subsoil and other constraints, sustainable crop and pasture rotations, and vegetation management have been the central research platforms across cropping, rangelands, and forestry ecosystems. Policy makers and politicians now nationally recognise that land degradation, carbon loss, and landscape degradation processes lead to deteriorating water quality, and that sustainable land, vegetation and water use, and greenhouse gas mitigation and climate change are the national and international priorities. In addition, he is a recipient of several awards and honours and also mentored 30 PhD scholars in the span of five decades and has published more than 450 publications with more than 23000 citations. He is a recipient of several awards/recognitions including the National Landcare Award, Hermitage (1994) and JA Prescott Award, Aus Soc

Soil Sci. (2008); *Professor* Dalal was made a member in the general division of the 'Order of Australia (AM)' in 2018 and also awarded 'Medal of Agriculture' by Ag Institute Australia in 2021 for outstanding service to science, and to farming through research and sustainable farming practices.

Dr Somasundaram Jayaraman graduated from the Faculty of Agriculture, Annamalai University (1995) and completed his post-graduation (1998) and Doctorate (2001) from Tamil Nadu Agricultural University (TNAU) in Coimbatore, Tamil Nadu, India. He is a Post-Doctorate Fellow (2015) at DSITI, Ecosciences Precinct, Brisbane, Australia. Earlier, he worked in the capacity of Senior Scientist and Principal Scientist at the Division of Soil Physics, ICAR–Indian Institute of Soil Science, Bhopal, Madhya Pradesh, India (2008-2023) and currently serving as Head, ICAR-IISWC, Research Centre, Udhagamandalam, Tamil Nadu (w.e.f. July 2023). He has more than two decades of experience in research and training in the field of natural resource management/conservation agriculture on soil properties and carbon sequestration vis-à-vis sustainable soil management. He has developed conservation agricultural practices for enhancing soil health and crop productivity in Western and Central India. In addition, farmers in rainfed regions of Central India were also made aware of soil health and developed low-cost farmer-friendly tools for assessing soil health in a participatory mode through field demonstrations. He has published more than 100 research papers in national and international peer-reviewed, high -impact journals with >2000 citations. In addition, he has authored/edited 10 books and published more than 50 book chapters. Besides, he has mentored post-graduate students (10), PhD scholars (3), and young scientists (5). He received two Gold Medals for PhD research work, Endeavour Research Fellowship in 2015 (by Australian Government), Australian Awards Ambassador in 2018 (by Australian High Commissioner to India), Scientist Award (2019), Leadership Award (2019), and Dr JSP Yadav Award for Excellence in Soil Science (2021). He is a Fellow of the National Academy of Biological Sciences (2022) and Fellow of the Indian Association of Soil and Water Conservationists (2022). He is Associate Editor of *Soil Research* (CSIRO) and Editor of *Farming System* and other national journals.

Introduction

1

Abstract

Fertile soils and human civilisations are closely interlinked. Worldwide, 4-4.8 billion hectares out of 13.3 billion hectares are used for crop and pastures, and these areas need to be protected from various kinds of land degradation. Fertility degraded soils lead to diminished current and potential capability to provide quantitative and qualitative goods and services because of one or more soil fertility degradative processes. These include soil erosion, nutrient removal in produce, plant biomass removal, oxidation (also anthropogenic burning), reduction and leaching, acidification, sodification, and structural degradation. Consequently, soil fertility degradation leads to organic matter loss, nutrient depletion, decline in nutrient bioavailability, soil physical degradation and biological degradation. It is imperative to identify these processes and devise sustainable management options for fertility restoration in the context of soil fertility-health-productivity-ecosystem services.

Key words

Land degradation · Soil fertility decline · Soil erosion · Aggregate breakdown · Biological degradation · Organic matter · Soil fertility restoration · Soil health

Soils are closely interlinked with human civilisation. Wherever good fertile soils exist in the adjoining river system, a greater human population exists, spreads, and prospers (Lal 2021a, b, 2023). Therefore, good fertile soil is fundamentally essential for sustainable and productive agriculture (Amundson et al. 2015). In the long term, agriculture is required to produce food of enough quantity and nutritionally good quality, for both human and livestock needs without impairing the fertility of the soil and the ecosystem functions and services (Dalal et al. 1997; Dalal 1998a, b; UN 2015). Soil fertility is thus defined as the ability of a given soil to support plant development and nutrition (Pauli 1967; Tisdale and Nelson 1975). On the other

R. C. Dalal, S. Jayaraman, *Soil Fertility Degradation, and Soil Health Restoration and Management*, https://doi.org/10.1007/978-981-96-0405-0_1

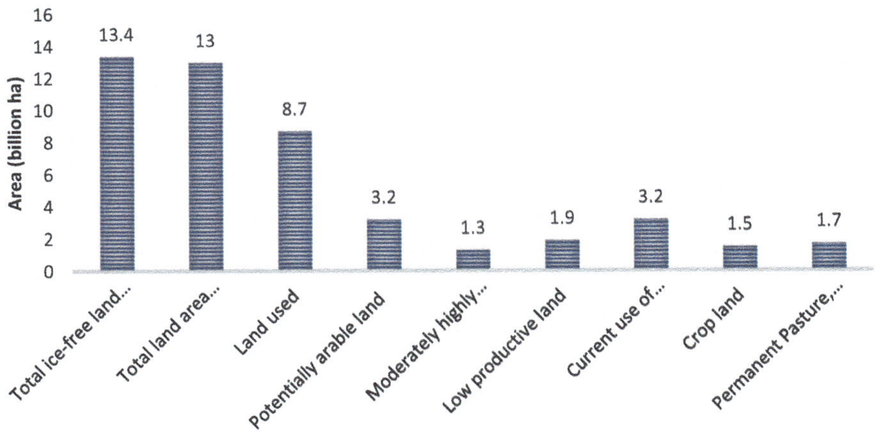

Fig. 1.1 Types and use of land area in the world. (Source: modified from Lambin and Meyfroidt 2021; Winkler et al. 2021)

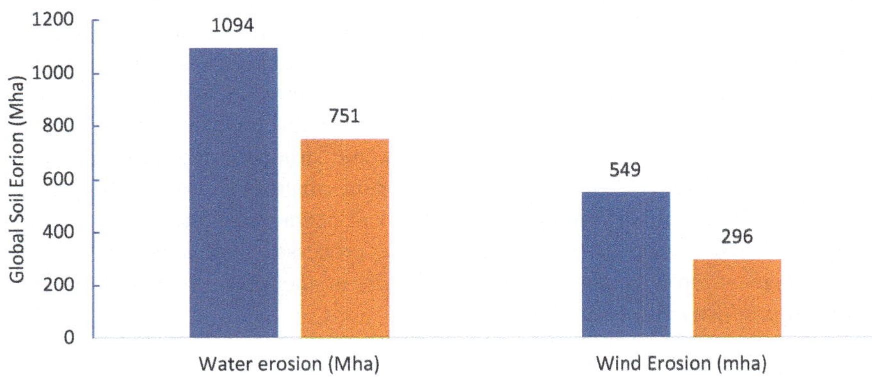

Fig. 1.2 Global soil erosion data (water and wind erosion). (Source: modified from Lal 2003)

hand, fertility-degraded soil is a degraded soil with its diminished current and/or potential capability to produce quantitative or qualitative goods or services as a result of one or more degradative processes (UNEP 1982; Dalal et al. 1997; Dalal 1998a, b; Tan et al. 2005). It follows, therefore, that agricultural productivity is unsustainable under continual degradation of soil fertility.

Worldwide, 4–4.8 billion hectares out of 13.3 billion hectares are used for crops and pastures (Lambin and Meyfroidt 2021; Winkler et al. 2021), and these areas need to be protected from various kinds of land degradation, as it poses a serious threat to food security (Lal et al. 1989; Lal 2009, 2015a, b) (Figs. 1.1 and 1.2). Land degradation not only deteriorates the land value (terrain deformation, soil loss via productivity loss) but adversely affects the socio-economic condition of the adjoining farming and rural communities (Barrow 1991; Lal 2003; Utuk and Daniel 2015). Soil erosion is a serious issue, as it degrades larger area in many countries

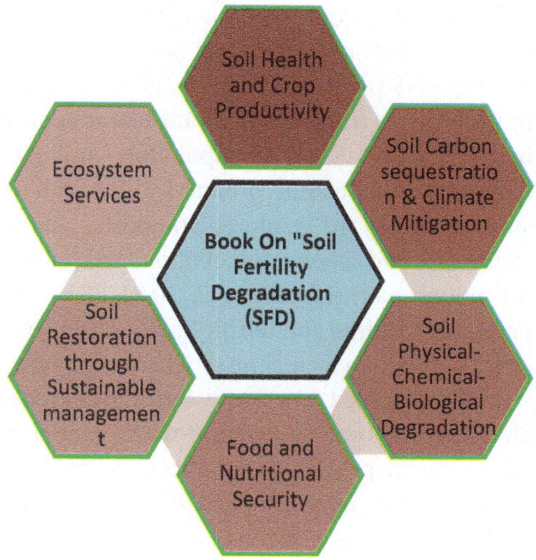

Fig. 1.3 An overview of soil fertility degradation

particularly in USA, 32% (Lal 2003); China, 30.7% (Guo et al. 2015); Africa, 16%; Europe, 17% (Lal 2003; Borrelli and Robinson 2017; Tamene et al. 2017); and Australia, 66% (ABARE 2009; Lumb, Land degradation fact sheet, https://www.australiancollaboration.com.au/). Similarly, in India, about 45% of the total land area has been already degraded by one or more soil degradation processes by a range of factors (NAAS 2010). In a recent estimate, FAO reported that annually seventy-five billion tonnes (Mg) of soil are eroded from arable lands globally, which has an equivalent monetary loss of US $400 billion annually (GSP 2017). Various kinds of soil/land degradation have been documented, namely, physical, chemical, biological degradation, and ecological degradation (Dalal and Bridge 1996; Lal 2015a, b), which have a severe impact on food production and sustainability.

Dominant processes of fertility degradation (diminution or depletion) include soil erosion, nutrient removal in produce, plant residue or biomass removal, oxidation (also burning), reduction and/or leaching, acidification, sodification and excessive element load, and aggregate breakdown or structural degradation (Dalal et al. 1997; Jie et al. 2002; Lal 2009; Maqsood et al. 2013) (Fig. 1.3). The nature of soil fertility degradation is the summative effects of one or more processes of fertility degradation, and it is reflected in organic matter loss, nutrient depletion, decline in nutrient bioavailability, soil physical degradation (led to poor soil structure, poor aeration/permeability, poor root anchorage), and biological degradation (decrease in activity and diversity of beneficial soil microflora and fauna and build-up in diseases, pests, and weeds) (Lal 2009). It is imperative to identify one or more processes involved in and affecting the nature of fertility degradation of a given soil so as to devise, promote, and implement sustainable management strategies that would maintain, restore, and, in the long term, enhance the fertility of such a soil (Dalal et al. 2021; Henry et al. 2023).

Management options for fertility restoration include management strategies based on soil (tillage and field/land configurations), plant (species and biomass management), soil–plant interaction (soil surface management), meso- and micro-biotechniques (rhizobium, arbuscular mycorrhizae and fungal inoculants, soil microbial RNA and DNA manipulation, and integrated pest management), inputs of natural and industrial products (lime and gypsum and biomass-based products, such as farmyard manure, composts, and sewage sludge; mined and synthetic fertilisers; and other products), and mechanical manipulation (such as deep ripping for plough-pan) (Davies et al. 2010, 2018). Not all management options are appropriate or required for fertility restoration. It is intended, therefore, to identify soil, crop, and/or environment(s) where one or a combination of management options would be needed to restore fertility-degraded soils (Dalal and Probert 1994; Lal 2015a, b). Moreover, all the management may not bring proper restoration, but management practices improve soil organic matter (SOM) or soil organic carbon (SOC), which may bring fertility restoration and enhancement in productivity as SOM acts as a central role in *soil-fertility-health-productivity-ecosystem services* (Oldeman 1997; Lorenz et al. 2019).

Scope of the Book

2

Abstract

This book covers the processes and nature of soil fertility degradation including erosion, physical, chemical and biological degradation, and management options for achieving soil fertility restoration including soil organic matter, productivity, environment and ecosystem services, and soil health and Sustainable Development Goals (SDGs). It is primarily restricted to the tropical and sub-tropical regions and soils of the orders, Oxisols, Alfisols, Ultisols, Inceptisols and Vertisols, used for dryland agricultural production, mostly in semi-arid and sub-humid climates. Also excluded from detailed consideration are metallic and synthetic organic pollutants and socio-economic aspects of fertility degradation and restoration of soils.

Key words

Soil fertility · Land degradation · Soil orders · Semi-arid region · Sub-humid region · Food · Fibre · Fodder · Energy security

This subject matter of the book is primarily restricted to the tropical and subtropical regions (approximately 35° north and south of the equator) and primarily soils of the orders, Oxisols, Alfisols, Ultisols, Inceptisols, and Vertisols, used for dryland agricultural production, mostly in semi-arid and sub-humid climates (approximately 3–8 wet months, mean annual precipitation of 250 mm to 2000 mm, and annual precipitation/potential evapotranspiration of >0.25, <0.75) (WRB 2014). Also excluded from the consideration of this review are metallic and synthetic organic pollutants and socio-economic aspects of fertility degradation and restoration of soils.

Excellent reviews have been published on a number of processes of fertility degradation, such as soil erosion (El-Swaify et al. 1983; Lal 1990; Gomiero 2016;

R. C. Dalal, S. Jayaraman, *Soil Fertility Degradation, and Soil Health Restoration and Management*, https://doi.org/10.1007/978-981-96-0405-0_2

Mosier et al. 2021), nutrient removal in produce and residues (Dalal and Probert 1994; Dickinson 2003; Maqsood et al. 2013), acidity (Sanchez et al. 1987; Butterly et al. 2022), sodification and salinisation (El-Swaify et al. 1983; Gupta and Abrol 1990), and aggregate breakdown (Lenvain et al. 1987; Dalal and Bridge 1996; Dalal and Jayaraman 2021). Some management options for fertility restoration, as outlined by Lal and Stewart (1990) and Lal (2015a, b), have been discussed in a number of proceedings on Vertisols (Latham 1987; Burford and Sahrawat 1989), Alfisols, Oxisols, and Ultisols (Latham 1987; Sanchez et al. 1987). In this book, discussions mainly will be based on the processes, nature, and consequences of soil fertility degradation in tropical and subtropical regions. This is justified in the belief that in these regions, improper management of soils for crop production leads to rapid fertility depletion and land degradation; consequently, droughts, famine, and hunger occur on a large scale and that optimum management of soils for crop production will sustain food and fibre production, maintain and/or restore soil fertility, and thus provide sufficient amounts of food, fibre, fodder, and energy for the human and livestock populations of these regions (Oldeman 1997; Bindraban et al. 2012).

Processes of Fertility Degradation

3

Abstract

Processes of soil fertility degradation described are soil erosion (water and wind erosion), nutrient removal in produce and plant residues, oxidation (including crop residue burning), reduction (including denitrification, waterlogging and anaerobiosis) and leaching, and loss of fauna, flora and microbial diversity. Acidification, salinisation and sodification, and soil aggregate breakdown in the semi-arid and sub-humid regions are the dominant processes of soil fertility degradation. Consequences and magnitude of soil fertility degradation including organic C and N losses and decline in productivity from these processes are considered and briefly potential management solutions suggested for restoration.

Key Words

Nutrient depletion · Water and wind erosion · Acidification · Salinisation · Sodification · Aggregate breakdown · Soil organic matter

Processes of soil fertility degradation considered in this chapter are those which are induced, modified, regulated, and initiated primarily by anthropogenic activity (Lal 2001, 2009; Mosier et al. 2021) or arising from unintended actions. The processes considered are soil erosion, nutrient removal in produce and plant residues, oxidation (including crop residue burning), reduction (including denitrification, waterlogging, and anaerobiosis) and leaching (Campbell 2017), and loss of fauna, flora, and microbial diversity.

R. C. Dalal, S. Jayaraman, *Soil Fertility Degradation, and Soil Health Restoration and Management*, https://doi.org/10.1007/978-981-96-0405-0_3

3.1 Soil Erosion

Soil erosion by water and/or wind is a geomorphic process, and in a natural ecosystem at a steady state, its degradative effect is negligible (Lal 2015a, 2015b) since it is in a steady state with replenishing the soil loss with continuing soil formation. When the natural ecosystem is disturbed by vegetation clearing (deforestation, burning, fuel wood removal, etc.) and/or land development and earthwork operations, soil erosion may become a most visible process of soil fertility degradation by en masse removal of generally the most fertile topsoil (Montgomery 2007; Mosier et al. 2021). In its less visible form but as much as or more insidious and detrimental effect on soil productivity, soil erosion selectively removes SOM (and with that especially N, P, and S) and fine soil particles and leaves behind coarse particles (Dalal et al. 2021). Briefly, water and wind erosion degrade soil fertility by increasing the losses of plant-nutrient elements, degrading soil structure (Lal 1998b, 2001), reducing plant available water capacity, exposure of sub-soil which is a poorer environment for soil microbial activity and crop growth, and increasing the non-uniformity in topography and soil characteristics (Williams et al. 1981; Lal 2009, 2015a, b). It also interacts with other fertility degradation processes, such as oxidation, reduction and leaching, acidification, sodification, aggregate breakdown, and biodiversity loss and eventually affects agricultural and environmental sustainability (Montgomery 2007; Lal 2015a, b).

3.1.1 Estimation of Soil Loss

3.1.1.1 Water Erosion

Annual average soil loss by water erosion can be estimated by three widely used models: the Universal Soil Loss Equation (USLE) (Wischmeier and Smith 1978; Wischmeier 1984), the Soil Loss Estimation for Southern Africa (Elwell 1981; Stocking 1981), and the FAO model of soil degradation by water erosion (FAO 1979); the latter two models are essentially similar to the USLE (Young 1989). All three models estimate soil loss using rainfall energy, soil erodibility, topography, and soil surface cover. Therefore, only the USLE and RUSLE models are presented in detail. The USLE model is

$$A = R.K.L.S.C.P. \tag{3.1}$$

where A = mean annual soil loss (Mg ha^{-1})
R = annual rainfall erosivity factor, $EI_{0.5}$ (Mg m ha^{-1})
K = soil erodibility factor (Mg ha^{-1} $EI_{0.5}{}^{-1}$)
L = slope length factor
S = slope steepness factor (LS, the topographic factor)
C = cover management factor
P = conservation practice factor

Rainfall erosivity, R, is the product of the rainfall energy and its maximum 0.5 h intensity for all rainfall events of more than 12.5 mm ($EI_{0.5}$ index). It is generally twice as high in the subhumid region compared to the semi-arid zone.

Rainfall erosivity in the USLE is based on mean annual or seasonal rainfall, and thus, it is less useful to predict soil erosion and runoff from individual rainfall events, especially from storms which are common in subhumid and semi-arid regions. Moreover, it assumes that runoff occurs uniformly in the entire catchment area (Renard et al. 1996; Luvia et al. 2022). This empirical equation does not represent the fundamental hydrological and erosion processes per se.

Onstad and Foster (1975) modified the rainfall erosivity factor in the USLE, taking into consideration both rainfall and runoff, as follows:

$$E = \left[0.646\ EI_{0.5} + 0.45\ (Rv)\ (Qp)^{0.333}\right]\ KLSCP \tag{3.2}$$

where E = soil loss (Mg ha^{-1}) from each erosion event
$EI_{0.5}$ = erosion index (Mg m ha^{-1})
Rv = runoff volume (m^3)
Qp = peak runoff rate (mm ha^{-1})

Soil loss is then the net soil loss from soil entrainment and deposition. Runoff volume (Rv) is predicted from daily rainfall (P, mm) and soil retentivity (S) from the relationship, $Rv = (P - 0.2\ S)^2/P + 0.8\ S$ (Knisel 1980), and runoff rate (Qp) from the relationship, Qp = P – I (Rose 1985) where I is the infiltration rate (I).

Based on hydrologic parameters, Rose (1985) accounted for the following three processes involved in net soil loss: rainfall detachment of soil, in which raindrops splash soil from the soil surface into the water of overland flow, sediment entrainment by overland flow or runoff detachment (no distinction is made between rill and interrill erosion), and sediment deposition, a process occurring continually owing to gravitational sediment settling velocity depending on sediment shape, size, and density. Since the first two processes increase the rate of soil loss and the last process decreases it, the net soil loss is determined by the relative magnitude of these rate processes, which are, to a significant extent, amenable to management practices. For example, sediment concentration in a runoff event is directly proportional to the fraction of soil surface exposed to overland flow or runoff detachment, as follows:

$$C = 2700\ S\ Cr\ h\ (1 - Wo/W) \tag{3.3}$$

where C = sediment concentration (kg m^{-3})
S = land slope, the sine of the angle of inclination
Cr = fraction of the soil surface exposed to overland flow
h = efficiency of net entrainment
W = stream power of overland flow (the rate of shear stress)
Wo = threshold value of stream power for the entrainment of soil or re-entrainment of deposited sediment process to commence (Rose 1985)

The efficiency of net entrainment, 'h', has been found to be inversely and non-linearly related to the surface cover (1 – Cr). The value of h fell from 0.7 to about 0.25 when surface cover increased from 0 (bare surface) to 0.1 for two Vertisols; that is, only a 10% surface cover reduced soil loss by almost threefold.

Soil erodibility, K, is a quantitative index of initially inherent soil susceptibility to erosion by water. Generally, the K factor (values in parentheses), that is, soil erodibility, increases from Ultisols to Alfisols: Ultisols (0.1) < Oxisols (0.2) < Inceptisols (0.23) < Vertisols (0.3) < Alfisols (0.35) (Wischmeier and Smith 1978; El-Saif and Dangler 1982). Besides soil texture and clay mineralogy, the K factor is a function of management responsive soil properties, such as SOM, aggregation, and dispersion (Young and Mutchler 1977), and consequently infiltration and permeability (Panagos et al. 2014; Raj et al. 2023). In general, a 10% reduction in SOM concentration is likely to increase its susceptibility to erosion by 10–25%, and a 25–50% loss in SOM increases its susceptibility to erosion by about 50% (Young 1989), while soil surface cover remains constant or decreases. In another study, ameliorating soils with farmyard manure (FYM) and rice straw or rotating perennial pasture with annual crops improved SOM and soil N status which further reduced soil erosion, runoff, and nutrient loss (Cogle et al. 2002).

The effects of the topographic factors, LS, slope length, and steepness, on soil loss were calculated by Wischmeier and Smith (1978) from the following relationships:

$$LS = (l/22.1)^{\mu}\left(65.41\ \sin^2 q + 4.56\ \sin\ q + 0.065\right) \qquad (3.4)$$

where l = slope length, m
q = angle of slope
μ = parameter dependent on slope (0.2 for slope <1% and 0.5 for slope 5%)

Equation 3.4 shows that twice the increase in slope length increases the rate of erosion by less than 50%, but that twice the increase in gradient more than doubles the rate of erosion. The topographic factor is a geomorphic feature of the landscape, but it can be modified by bench terracing (using engineering measures as well as vegetative measures) and other mechanical manipulations (e.g. contour/graded bunding, continuous contour/staggered trenching). The effect of the topographic factor on soil erosion can be reduced by both soil management as well as plant management and mechanical manipulations (Fig. 3.1, 3.2, 3.3, 3.4 and 3.5), such as contours, bays, and ditches, although the latter by themselves may not necessarily be effective (Freebairn and Wicker 1986).

The cover management factor, C, is the ratio of soil loss from land under specified crop and/or residue cover to that from tilled bare fallow. The C factor varies from 1.0 for bare fallow, 0.4 for a moderately yielding cereal crop, 0.04 at 100% cover, 0.01 for a good pasture, to 0.001 for dense forest (Roose 1986). Generally, the C factor in subhumid regions (0.6) exceeds those in semi-arid regions (0.3) (FAO 1979). In both these regions, especially in the latter, inadequate soil moisture and periodic droughts reduce the periods when soil cover is provided by the crop and also limit the amount

Fig. 3.1 Topography factor and along-the-slope cultivation impacts soil erosion. (Source: Jayaraman and Dalal)

Fig. 3.2 Poorly managed land configuration/conservation measures are leading to soil erosion and degradation. (Source: Jayaraman and Dalal)

Fig. 3.3 Crop management and crop cover impact soil erosion. (Source: Jayaraman and Dalal)

of crop residue produced. Since pasture establishment and persistence are commonly constrained by available soil moisture, proper management of available crop residues is essential for reducing soil loss (Fig. 3.6). Thus, in two Vertisols (Udic Pellustert and Typic Chromustert) in the semi-arid subtropics, only 10% and 50% covers provided by wheat crop residues reduced the soil loss by more than 60% and 90%, respectively (Freebairn et al. 1991) (Fig. 3.6).

Similarly, Montgomery (2007) reported that land use had a conspicuous effect on the rate of soil erosion. It varied from 0.053 mm year^{-1} (native vegetation) to 0.124 mm year^{-1} (conservation agriculture) to 3.939 mm year^{-1} (conventional farming). Therefore, choosing appropriate farming practices (both agronomic and soil management) has a significant impact on soil erosion (Singh et al. 2020).

The conservation practice factor, P, is the ratio of soil loss with a specific conservation practice to that with up-and-down slope cultivation (Wischmeier and Smith 1978). The most important of these conservation practices are contouring (most effective on 3–12% slopes and 40–100 m slope lengths), strip cropping (pasture-cereal rotations in strips), contour strip cropping (especially with crop residue cover, it is very effective), terracing (broad base on gently sloping lands, backslope (inward) terrace on steeper lands), diversions, graded furrows, and drainage. The conservation support practices along with surface cover reduce not only the soil movement within a field but also net sediment loss (increased deposition) and runoff from the field (Fig. 3.4). Reductions in off-site sediment by these practices are

Fig. 3.4 Crop cover and residue retention play a significant role in reducing soil erosion. (Source: Jayaraman and Dalal)

reduced by as much as 95% (Wischmeier and Smith 1978; Nelson 1997). For example, Cogle et al. (2011) reported from a 3-year study that soil loss ranged from 30 Mg ha^{-1} under bare conditions (without cover or crop) to <6 Mg ha^{-1} under cropping in Red Ferrosols in tropical northern Australia.

Revised Universal Soil Loss Equation (RUSLE)

The Revised Universal Soil Loss Equation (RUSLE) is an upgrade of USLE, which is land use independent and uses the same empirical equation (Luvia et al. 2022). It can be used on a wide range of land use such as cropland, disturbed forestland, rangeland, construction sites, mined land, reclaimed land, landfills, waste disposal sites, and other lands where rainfall and its associated overland flow cause soil

Fig. 3.5 Runoff collecting device installed to monitor soil and nutrient loss. (Source: Jayaraman and Singh)

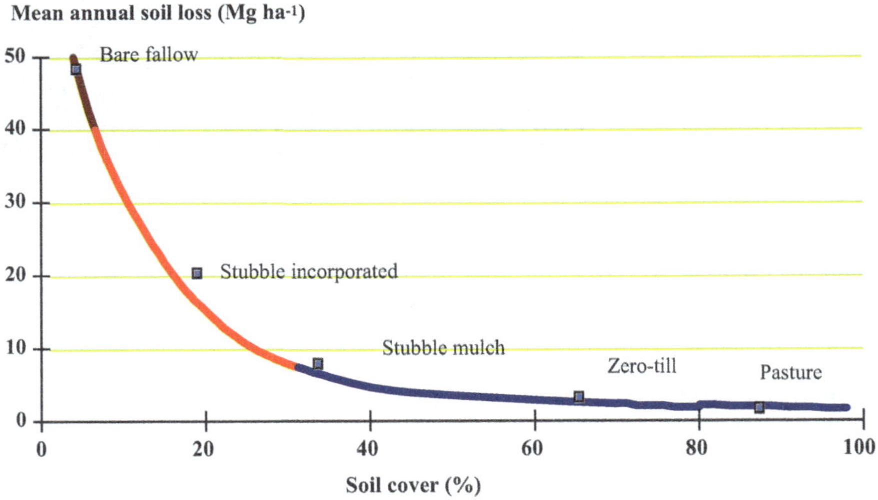

Fig. 3.6 Relationship between hill-slope water erosion and soil surface cover under four summer fallow tillage practices and a pasture on a Vertisol (Typic Pellusterts) in semi-arid (annual rainfall, 735 mm) subtropical Australia. (Supplied courtesy of Dr. David Freebairn 2004)

erosion. Unlike USLE, RUSLE better predicts situations where long-duration rainfall can saturate the soil, leading to reduced intake and greater erosional runoff. In contrast with USLE, RUSLE can handle converging and diverging terrain as well as considers areas with net sedimentation (Renard et al. 1991, 1996; Yoder and Lown 1995).

Advantages of the RUSLE Model

(i) Includes corrections for high 'R'-factor region with flat slopes in order to predict the splash erosion.
(ii) It helps in the development of a seasonally variable 'K' factor (soil erodibility).
(iii) It helps in calculating LS products under slopes of varying shapes and also predicts LS under rill to interrill erosion.

3.1.1.2 Wind Erosion

Soil loss by wind erosion occurs in the semi-arid tropical and subtropical regions (Lal 1990a), but it can also occur in the subhumid regions with pronounced long dry seasons and variable rainfall (WMO 1983; Fryrear et al. 1999). Wind erosion occurs by unique processes, namely, suspension, saltation, and surface creep within 2 m of the soil surface.

The wind erosion equation (Woodruff and Siddoway 1965) is given below, but the basic technology has been described as the Universal Wind Erosion Equation (WEQ) (USDA 1961). The main cornerstone of WEQ is the soil erodibility factor (I), whereas the foundation of the revised wind erosion equation is the wind characteristics:

$$WE = I.WK.WC.WL.VE \tag{3.5}$$

where WE = wind erosion (Mg ha^{-1})
I = soil erodibility index (soil erodibility and knoll erodibility)
WK = soil-ridge roughness factor
WC = climatic factor (wind velocity and surface soil moisture)
WL = field length along the prevailing wind direction (distance across the field and unsheltered distance, m)
VE = equivalent quantity of vegetative cover (quantity of vegetative cover, in kg ha^{-1}, kind of vegetative cover and orientation of vegetative cover)

Equation 3.5 is used to predict average annual soil loss by wind erosion. Cole et al. (1983) and Williams and Renard (1985) modified Equation 3.5 to predict daily wind erosion by adjusting WK, WL, and VE on a daily basis. For example, the vegetative cover equivalent factor is estimated daily as a function of standing live crop biomass, standing dead residue, and flat crop residue:

$$VE = 0.2533\ (g_1 BIOM + g_2 SR + g_3 FR)^{1.363} \tag{3.6}$$

where g_1, g_2, and g_3 = crop-specific coefficients
BIOM = above-ground biomass of a growing crop (kg ha^{-1})

SR = standing residue from the previous crop (kg ha^{-1})
FR = flattened crop residue at the soil surface (kg ha^{-1})

Also, the finer the crop residue and the greater its surface area, the more it reduces the wind velocity, and hence, it is more effective in reducing soil erosion (Woodruff and Siddoway 1965) since the rate of soil movement varies directly with the cube of wind velocity. Thus, the effectiveness of flattened wheat residue may be only about 50% of that of standing residue (Unger and Stewart 1988).

In the late 1990s, the wind erosion equation was revised by Fryrear et al. (1999), where RWEQ is the transport mass, i.e. soil mass transported by the wind in the band of unit width to a specified height of 2 m from the soil surface, but WEQ does not specify the height. To overcome some concerns/deficiencies in using WEQ and RWEQ, the Wind Erosion Prediction System (WEPS) has been launched by USDA (Wagner 2013). However, it created a spark in using conventional field experiments along with a wind tunnel approach in deciphering the soil properties and its susceptibility to wind erosion.

3.1.2 Magnitude of Soil Loss

Natural soil erosion is a geomorphic process; however, it adversely affects food production and sustainability. It is generally believed that the rate of natural soil erosion loss approximates the rate of soil formation in a steady-state ecosystem (Montgomery 2007). Kadomura and Yamamoto (1978) estimated that the rates of natural soil erosion vary from 0.1 to 7 Mg ha^{-1} $year^{-1}$ for the terrestrial ecosystem, depending upon the relative magnitude of factors affecting soil erosion (Eqs. 3.1 and 3.5). A generalised relationship between mean annual precipitation and soil loss by wind and water erosion, depending upon natural vegetation cover, shows that a varied but small loss of soil is likely, especially in the semi-arid regions (Fig. 3.3 and 3.4). However, once a steady-state ecosystem is perturbed, especially by removing vegetation cover and creating soil disturbance, soil erosion increases manifold (Lal 2003; Montgomery 2007; Borrelli et al. 2017) (Fig. 3.3 and 3.4), thus resulting in rapid degradation of soil fertility (Dalal 2021). In general, the relative impact of erosion on soil fertility by wind erosion decreases and that by water erosion increases with increasing rainfall (Marshall 1973; Montgomery 2007).

3.1.2.1 Water Erosion

The extent or rate of soil erosion is influenced by many factors, viz., parent material, soil texture, organic matter content, topography (slope length and steepness), plant cover, land use, tillage practices, and climate (Borrelli et al. 2017; Guo et al. 2019).

Lal (1991) has pointed out the difficulty of assessing the information on soil loss by water erosion because it has not always been stated in the reports whether it was soil movement, soil loss, or sediment loss, since the latter is a function of sediment delivery ratio [the ratio of sediment delivered at a given location to the gross soil erosion (movement) from the given area at that location] which is generally two to five times less than the soil movement.

In one of the early studies conducted, East (1985) reported that soil losses of as high as 405 Mg ha^{-1} under conventionally cultivated areas and unprotected land of Atherton and Herberton Shires of Australia. Dregne (1982) and Lal (1991) have comprehensively reviewed the soil loss by water erosion. Depending upon the soil erosion factors, in the semi-arid and subhumid regions, soil erosion in cultivated Ultisols, Oxisols, Alfisols, Vertisols, and Inceptisols is estimated to be 5, 10, 20, 40, and 80 Mg ha^{-1} $year^{-1}$ (Narayana and Sastry 1985; Bhattacharya et al. 2007; Sharda et al. 2013), although soil loss up to 6000 Mg ha^{-1} $year^{-1}$ has been recorded in Burundi (Van Lancker and Bruggeman 1982). Sharda and co-workers (2013) estimated that 91% area of the Indian sub-continent has an erosion rate ranging from 5 to 40 Mg ha^{-1} $year^{-1}$, which needs immediate interventions of soil and water conservation practices. Another important aspect of soil erosion processes is runoff water loss which varies from 10% to 60% of annual rainfall. This process not only deprives crops of available water but also carries with it fine sediments, nutrients, micro- and mesofauna, and added contaminants to the downstream water resources such as lakes, reservoirs, ponds, rivers, and ultimately to the sea.

Montgomery (2007) clearly stated that conventionally tilled fields recorded higher erosion rates (i.e. one to two times) greater than the rate of soil formation in comparison to native vegetation as well as long-term geological erosion. However, no-till farming recorded an erosion rate much closer to the rate of soil formation. Different researchers estimated global geological erosion rates (0.016 to 0.024 mm $year^{-1}$) and soil formation rates (0.053 to 0.083 mm $year^{-1}$) (Fig. 3.7).

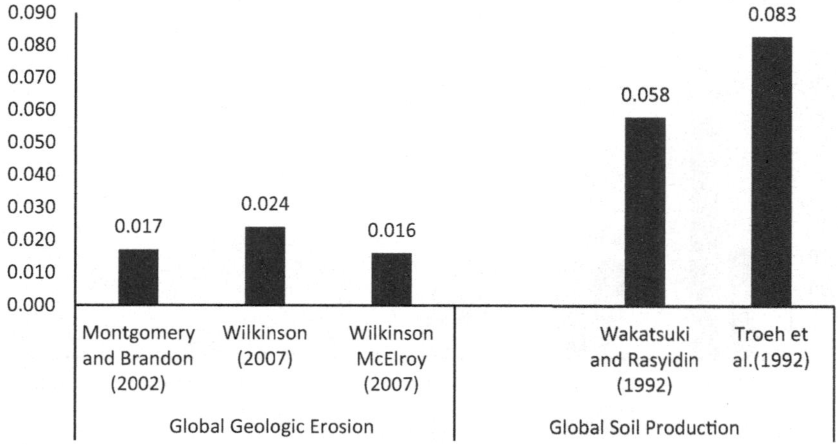

Fig. 3.7 Globale geologic soil erosion versus soil production/formation rate (mm $year^{-1}$). (Source: Modified from Montgomery 2007)

3.1.2.2 Wind Erosion

Wind erosion is a significant process of soil loss and hence of fertility degradation and productivity deterioration, which mainly occurs in arid and semi-arid regions. Wind erosion also inflicts air pollution and causes health hazards to human and livestock populations. Quantitative estimates of soil loss due to wind erosion are much less available than those due to water erosion (Dregne 1982), although its contribution to soil fertility degradation on Alfisols and Inceptisols could be considerable. Williams and Renard (1985) estimated soil loss by wind erosion to be 1–14 (wheat field), 14–16 (grain sorghum), and 65–717 Mg ha^{-1} $year^{-1}$ (cotton) in Texas, USA.

It is always challenging to accurately estimate/predict wind erosion in an area (from local scale to large scale) despite the availability of many empirical equations such as WEQ, RWEQ, and ^{137}Cs tracing technique, as well as inserted drill rods at fields, and models, namely, Wind Erosion Prediction System (WEPS) and Wind Erosion Stochastic Simulator (WESS). In order to overcome some limitations in other methods and models, Yang et al. (2022) predicted global wind erosion for the period of 2001–2010 using a distributed version of the revised wind erosion equation model (DRWEQ) (Fig. 3.8). Africa and Asia accounted for 91% of the global total soil loss, which represents 62% of the global wind erosion area; encouragingly, wind erosion was under decreasing trend during the 2001–2010 period.

3.1.3 Magnitude of Soil Fertility Degradation

Water and wind erosion degrade soil fertility by removing soil en masse, thus reducing the rooting depth, plant available water capacity, and soil structure, exposing poorer sub-soil, and causing losses of plant nutrient elements, microbial biomass, and diversity, thereby further exacerbating land degradation (Barrow 1991; Oldeman 1997; Lal 2003, 2009; Adimassu et al. 2019).

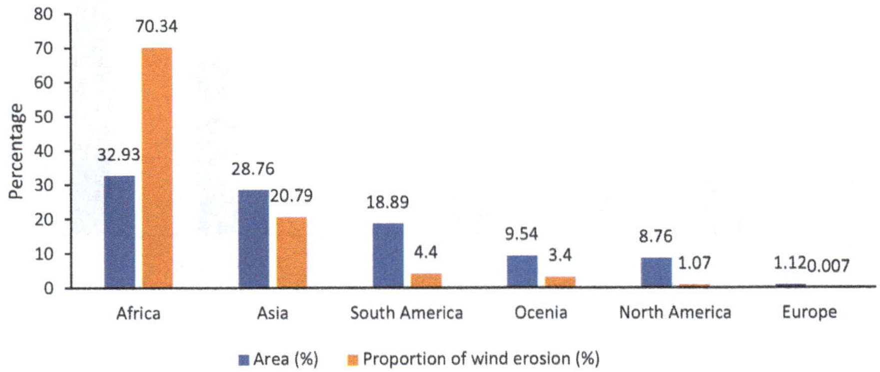

Fig. 3.8 Comparison of wind erosion in different continents. (Source: Modified from Yang et al. 2022)

Table 3.1 Estimated sediment and nutrient loss from different continents

Continent	Area[a] (106 ha)	Sediment loss[a] (Mg ha^{-1} year^{-1})	Nutrient loss in sediments[b] (kg ha^{-1} year^{-1}) N	P	K	Total nutrient loss in sediments (x10^{6}kg year^{-1}) N	P	K
Africa	2 981	0.47	0.71	0.47	0.94	2 117	1	2 802
Asia	4 489	1.66	2.49	1.66	3.32	11	401	14
Australia	796	0.32	0.48	0.32	0.64	178	7	903
Europe	967	0.43	0.65	0.43	0.86	382	452	509
North and Central	2 044	0.73	1.10	0.73	1.46	629	255	832
America	1 798	0.93	1.40	0.93	1.86	2 248	416	2 984
South America						2 517	1	3 344
							492	
							1	
							672	
Total	13 075	0.97	1.46[c]	0.97	1.94	19 071C	12 688	25 374

[a]Adapted from El-Swaify et al. (1982)
[b]Assuming N, P and K concentrations of the topsoil layer, respectively, 0.075 %, 0.05% and 0.1%; enrichment ratio of sediment as 2 (Knisel 1980)
[c]Total and rate of N loss from the land surface are estimated to be 31 000 × 10^6 kg year^{-1} and 2.5 kg ha^{-1} year^{-1}, respectively (Post et al. 1985)

Estimates of N, P, and K losses through the loss of sediments from various continents are 1.5, 1, and 2 kg ha^{-1} year^{-1}, respectively (Table 3.1). A conservative estimate of rates of loss of nutrients from cultivated lands (1461 × 10^6 ha, Dregne 1982), assuming 50% loss from this source and sediment delivery ratio of 0.5, is about 13, 8, and 17 kg ha^{-1} year^{-1} of N, P, and K, respectively. Erosion losses of plant nutrient elements, N, P, and K from cultivated Vertisols and Inceptisols, generally vary from 10 to 250 kg N ha^{-1} year^{-1}, 1 to 20 kg P ha^{-1} year^{-1}, and 5 to 200 kg K ha^{-1} year^{-1} in eastern Australia; the losses are much smaller from soils under pasture or native vegetation cover. Cogle et al. (2011) reported nutrient losses of 35 kg N and 35 kg P ha^{-1} year^{-1} under bare soil conditions and <2 Mg ha^{-1} year^{-1} of soil loss under cropping (peanut-maize) from Red Ferrosols in northern Australia. Most of this loss of N (usually >90%) occurs as organic N in eroded sediments and suspensions. Usually, only 1–2% of soil N occurs as mineral N, which is susceptible to loss in soil solution by leaching, runoff, denitrification, or volatilisation. Almost all of the loss of P (>95%) and that of K (>90%) also occur in eroded sediments. Hence, the loss of sediment by erosion from soil leaves it with less N, P, and K and potentially lower productivity (i.e. erosion induced lower productivity) (Table 3.2). Thus, soil erosion is an important and insidious process of long-term fertility degradation of soils (Gantzer et al. 1990; Campbell 2017; Mosier et al. 2021; Yang et al. 2022).

Since SOM and its associated N and P contents as well as available and potentially available plant nutrient elements typically increase rapidly towards the soil surface (within soil orders: Ultisols > Oxisols > Alfisols > Inceptisols > Vertisols), these soil constituents are highly prone to soil erosion. Furthermore, SOM and its

Table 3.2 Organic C, total N, P and K in noneroded and corresponding eroded sediment

	O.C (%)		Total N (%)		Total P (%)		Total K (%)	
Soil order	Noneroded	Eroded	Noneroded	Eroded	Noneroded	Eroded	Noneroded	Eroded
Oxisol[a]	2.59	1.40	–	–	–	–	0.12	0.06
Oxisol[b]	2.02	0.87	0.167	0.080	–	–	–	–
Ultisol[b]	2.20	0.78	0.173	0.071	–	–	–	–
Ultisol[c]	5.50	2.40	0.480	0.100	0.10	0.06	0.70	0.23
Alfisol[c]	5.50	2.65	0.480	0.200	0.10	0.06	0.70	0.33

Source: [a]Moberg (1972), [b]Wen and Lin (1986), [c]Vy and Thi (1989)

associated constituents have a high enrichment ratio in the eroded sediment, so the loss of SOM and plant nutrient elements is exaggerated relative to the amount of total soil loss (Roose 1977; Lal 1980, 1983, 1985, 1990, 1991, 1996; Sharpley 1985; Rose and Dalal 1988a, b; Ghadiri and Rose 1991; Cogle et al. 2011).

The soil erosion processes responsible for the sediment enrichment ratio (i.e. the ratio of the concentration of a constituent of eroded sediment to its concentration in the bulk soil) are believed to be rainfall detachment, wherein peeling off of outer enriched layers of aggregates occurs due to raindrop impact, that is, *raindrop stripping* (Ghadiri and Rose 1991; Bot and Benites 2005; Cogle et al. 2002) and *deposition* (Foster et al. 1985; Rose and Dalal 1988a, b) since deposition is a highly selective process due to differences in particle and aggregate size, shape, and density and hence different settling velocity. The relative contribution of either or both of these mechanisms of enrichment ratio in soil erosion processes depends on the degree of aggregation of the sediment (Bot and Benites 2005). Raindrop stripping may be involved in the selective removal of sorbed chemicals and organic matter and its constituents from the surfaces of relatively stable aggregates (Bot and Benites 2005: Wang et al. 2019a, b). Provided that raindrop stripping is not a main contributing factor towards the enrichment of the eroded sediments, that is, water depth of more than a few mm (Ghadiri and Rose 1991), then the enrichment of the eroded sediment is relatively small from well aggregated fine-textured soils. Moreover, clay is evenly distributed in aggregates and eroded sediments in these soils. Even if the latter becomes further enriched in clay, it may not necessarily be enriched in organic matter and its constituents (Rose and Dalal 1988a, b). On the other hand, fine particles in the eroded sediments, such as silt and clay, detached from poorly aggregated coarse-textured (sandy) soil remain in suspension for a much longer period than sand, and thus, such sediments become enriched in fine particles and associated organic matter and sorbed constituents. Wang et al. (2019a, b) reported that particle size distribution (PSD) of the sediment is a vital factor, which contributes 50% of the transportation of SOC and soil inorganic carbon (SIC) during erosion of sediments.

The enrichment ratio of the eroded sediment depends on the sediment delivery ratio, sediment characteristics, and the relative contribution of interrill and rill erosion to sediment yield (Knisel 1980; Rose and Dalal 1988a, b; Li et al. 2021a, b). For example, if only 10% of the sediment is transported from a field, that is, the sediment delivery ratio is 0.1, then the enrichment ratios of the eroded sediment with respect to organic matter originating from sandy soil and clay soil are 5 or more and 1.5 or less, respectively (Fig. 3.9). It follows, therefore, that the rate of fertility degradation will be much faster in a coarse-textured soil than in a fine-textured soil at a similar sediment delivery ratio and sediment loss (Al-Kaisi 2001). Nathan et al. (2022) reported that the combined application of soil and water conservation measures with nutrient application helped in the reduction of nutrient and SOM loss, ecosystem disturbances, and monetary losses. They reported that all the treatments had an enrichment ratio (ER) of more than unity, indicating more ecosystem disturbances (Bertol et al. 2017).

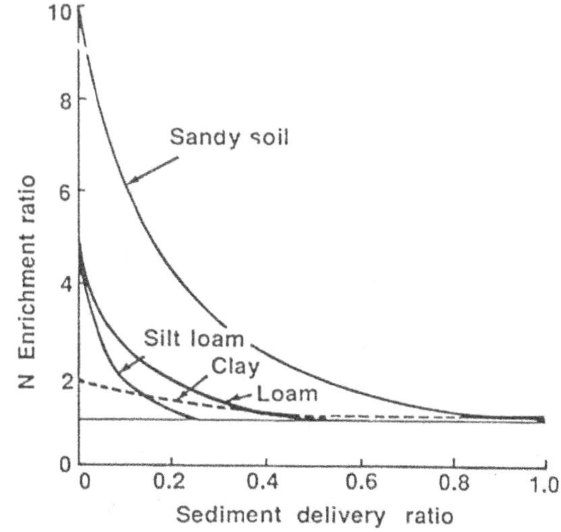

Fig. 3.9 Relationship between sediment delivery ratio and estimated N enrichment ratio from soils of different textures. (Source: Rose and Dalal 1988a, b)

The enrichment ratio (ER) declines as the sediment yield increases. The relationship between ER and sediment yield is satisfactorily expressed by Equation 3.7:

$$\text{Ln (ER)} = A - B \ \ln \text{ (SED)} \tag{3.7}$$

where SED = accumulated sediment (soil) loss (kg ha^{-1})
A and B = constants which are generally in the range of $1 < A < 3$ (2.13 ± 0.65 for total N and 1.84 ± 0.62 for total P) and $-0.3 < B < -0.1$ (-0.26 ± 0.14 for total N and -0.16 ± 0.12 for total P)

For a course-textured Alfisol (0.16 kg clay kg^{-1} soil), the A values were found to be 2.4 for organic C, 2.6 for total N, 2.3 for organic P, and 1.8 for $NaHCO_3$ extractable P; the corresponding B values were −0.24, −0.26, −0.24, and −0.19 (Dalal and Loch 1984), which are within the range measured in a number of soils (Menzel 1980; Sharpley 1980; Ghadiri and Rose 1991).

The dominant erosion processes (i.e. detachment, entrainment, transport, and deposition) differ between interrill and rill erosion, such that the eroded sediment from interrill erosion may be enriched by two to three times compared with that from the latter (Rose and Dalal 1988a, b). This is mainly because in interrill erosion there are greater opportunities for differential settling velocities of aggregates (and particles) (Foster et al. 1985), turbulence within the flow layer due to raindrop impact (Dalal and Loch 1984; Schiettecatte et al. 2008a), raindrop stripping (Ghadiri and Rose 1991), and sorting of aggregates during entrainment and re-entrainment (Rose and Dalal 1988a, b). Thus, nutrient element loss would be proportionately more than soil loss when interrill erosion is the dominant soil erosion process, that is, in small runoff events of short duration, on low slopes (such as in Vertisols), or small field plots. Similar to the improved erosion equations based on separate terms for interrill and rill erosion processes, organic matter and associated nutrient element loss would be better estimated when these are based on separate terms for the two processes.

The sorting of aggregates (and particles) eroded by wind primarily takes place in the suspension process because the smaller aggregates, usually enriched in fine particles, organic matter, and associated constituents, remain longer in suspension resulting in greater particle selectivity and higher enrichment ratio (Hagen and Lyles 1985; Schiettecatte et al. 2008a). Schiettecatte et al. (2008b) studied the detachment capacity (Dc) and transportation capacity (Tc) through rainfall simulations and reported that detachment by runoff was directly linked to total runoff discharge. The Dc values were four to 50 times lower than the Tc with respect to high and low runoff discharge. This finding suggested that detachment by runoff represents only part of the transported sediment.

Organic matter and nutrient loss by soil erosion are generally estimated as the product of total sediment loss and nutrient element concentration in the bulk soil (Schiettecatte et al. 2008a; Wang et al. 2019a, b). The nutrient element loss and hence potential loss of soil productivity would be considerably underestimated if enrichment ratios of eroded sediments are ignored. Unfortunately, the management practices used to reduce soil erosion do not proportionately reduce the loss of organic matter, associated nutrient elements, and chemicals from contaminated soils. Research is required to identify the management practices that not only reduce soil loss but also reduce enrichment ratios of the eroded soil (Schiettecatte et al. 2008b; Li et al. 2021a, b).

3.1.4 Soil Erosion and Crop Productivity

Soil erosional process significantly alters the soil physical, chemical, and biological properties, and these directly influence crop growth and productivity (Lal and Moldenhauer 1987; Lal 2003; Bakker et al. 2004). The crop productivity under eroded fields is much lower than the uneroded fields due to the removal of organic carbon and essential plant nutrients.

The effects of soil erosion on crop productivity have been adduced from a decline in crop yields (Lal 1985, 1990, 1991) by (i) desurfacing, (ii) measured soil loss by erosion, and (iii) reduced soil depth (Lal 1983, 1990, 1991; Aveyard 1983; Castillo and Paez 1989; Gantzer et al. 1990; Lal 1998; Arriaga and Lowery 2003; Bakker et al. 2004; Sharda and Dogra 2013). The limitations of each of these approaches and cautions in interpreting the results from such studies have been pointed out by Meyer et al. (1985). Similarly, Bakker et al. (2004) pointed out that experimental methodologies have a profound influence on estimating the effect of soil erosion on productivity loss from different experiments. They reported that comparative-plot techniques registered an average reduction in crop productivity by 4.3% per 10 cm of soil loss whereas the transect method by 10.9% and desurfacing experiments by 26.6%. This implies that the comparative-plot method proved realistic in estimating productivity loss compared to the desurfacing and transect method.

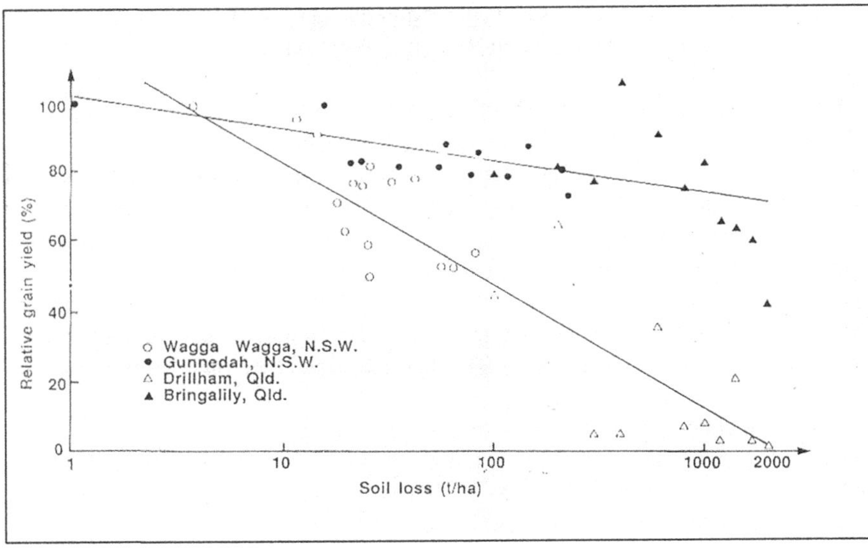

Fig. 3.10 Relationship between soil loss and relative wheat grain yield under dryland conditions in eastern Australia. Circles represent cumulative soil loss under natural rainfall over 26 years (Gunnedah) and 30 years (Wagga Wagga) followed by two and four wheat crops, respectively (from Aveyard 1983). Triangles represent soil removal by desurfacing, followed by a wheat crop (P.J. White, unpublished data). Soil profiles at Gunnedah and Bringalily are medium to deep, whereas the others are shallow, with either texture-contrast (Wagga Wagga) or saline sub-soil (Drillham). m_ Rl yield (%) = 125 – 16.6 ln soil loss (Mg ha^{-1}) $r^2 = 0.85$, •Ù Rl yield (%) = 103 – 4.6 ln soil loss; $r^2 = 0.36$. (Source: Collated and developed by R.C. Dalal)

The effects of soil erosion by desurfacing and natural soil loss on grain yield in an Alfisol and three different Vertisols were reported by Rose and Dalal (1988a, b) from the semi-arid regions of Australia (Fig. 3.10). The rate of decline in yield was more moderate for Vertisols of medium to deep profiles than for an Alfisol with a texture-contrast profile or a Vertisol with a saline sub-soil. The moderate decline in productivity was corrected by fertiliser application and irrigation, a result similar to that on Houston black clay Vertisol (Baird and Knisel 1971), but not that on the Vertisol with saline sub-soil. Similar results have been reported from subhumid and semi-arid regions on Ultisols (Larson et al. 1985; Mbagwu 1988, 1989), Alfisols and Oxisols (Lal 1983; El-Swaify and Dangler 1982; El-Swaify et al. 1982; Lo 1989), Mollisols (Lindstrom et al. 1986), Inceptisols and Vertisols (Freebairn et al. 1991), and loess soil (Schumacher et al. 1994). In a subtropical country like India, Sharda and Dogra (2013) reported that productivity loss due to soil erosion in rainfed cereals ranged from 0.02–1.09, 0.01–0.63, and 0.004–0.44 Mg ha^{-1} for cereals, oilseeds, and pulses, respectively, with mean of 0.16 Mg ha^{-1} (= Rs 2484/ha based on the minimum support price during 2011–2012) in productivity of rainfed crops.

Erosion-induced loss in crop productivity is soil-specific, and it is also affected by environment and management (Stocking and Peake 1986). The following general conclusions can be drawn from soil erosion-crop productivity studies:

1. Yield and erosion are related either linearly or squarely to the soil loss for relatively young soils with appreciable reserves of weatherable minerals throughout the profile or deep Vertisols with less marked gradations in soil properties with depth. But these are strongly related either exponentially (Stocking and Peake 1986) or log-normally (Rose and Dalal 1988a, b) for soils with marked concentration gradient with depth in organic matter and associated nutrient elements (Ultisols), texture-contrast profile (Alfisols), sub-soil acidity and low activity clays (Ultisols and Oxisols), and sub-soil salinity and sodicity (Vertisols).
2. There is growing evidence of a strong correlation between reduced crop yields and decreased SOM or SOC stocks in eroded soils. Lindstrom et al. (1986) reported decreased corn grain and stover yield due to topsoil removal and subsequent reduction in SOC level in the Ap horizon. Similarly, Schumacher et al. (1994) reported a reduction in corn yields under eroded soil of 11 sites in the North Central region of the USA. In another study conducted by Lobo et al. (2005) located in Chaguaramas in the Central Plains of Venezuela, four different study locations with the same soil type (with slopes ranging from 3% to 6%) coupled with different levels of erosion were selected: Chaguaramas I (slightly eroded), Chaguaramas II (moderately eroded), Chaguaramas III (moderately eroded), and Chaguaramas IV (severely eroded). They reported that the productivity index (PI) decreased with a higher rate of soil erosion; it was largely influenced by changes in available water storage capacity, bulk density, and pH. Similarly, the erosion risk was strongly influenced by slope gradient and rainfall intensity.
3. The impact of a unit loss of soil on yield in subhumid and semi-arid, tropical, and subtropical Alfisols and Ultisols appears to be an order of magnitude greater than on the equivalent temperate soils. Since soil erosion hazards are higher in the former regions, annual declines in productivity by erosion could easily be twenty or more times more serious than that in the temperate climate (Larson et al. 1985).
4. Overall changes in physical, chemical, and biological properties contribute to soil fertility degradation and loss in productivity (Lal 1998; Arriaga and Lowery 2003). These include loss in SOM and consequent loss in CEC and nutrient retention, especially in Ultisols and Oxisols; nutrient losses from topsoil; reduction in plant available water capacity, especially in semi-arid and seasonally dry regions; losses in structural stability and compaction; acidification and increase in exchangeable aluminium in Ultisols and Oxisols; and increasing salinisation and sodification in Vertisols and loss of microbial biomass and biodiversity in all soils.

Some of these factors relating to soil erosion and crop productivity are expressed by Pierce et al. (1983) in the productivity index (PI) for physical effects as follows:

$$\mathrm{PI} = \mathrm{S}_{=1}\ (A_i.B_i.C_i.\mathrm{WF}_i) \tag{3.8}$$

$$PI = \sum_{i=1}^{n} (A_i * C_i * D_i * E_i * RI_i),$$

where A_i = sufficiency of available water capacity
B_i = sufficiency of bulk density (or permeability to air and water)
C_i = sufficiency of pH
RI or WF = root weighing factor
r = number of 0.1–m increments in the root depth or n = number of soil layers of the root zone depth

Equation 3.8 can be modified to include nutrient loss and biological degradation (e.g. build-up of plant pathogens in soil, loss of arbuscular mycorrhiza fungi, AMF):

$$PR = PI.Ns.Bs.Dr \quad (3.9)$$

where PR = crop productivity rate per year
PI = productivity index for physical effects
Ns = integrated nutrient element sufficiency
Bs = integrated biological sufficiency
Dr = productivity decrease due to excess element load and/or organic pollutants

Considerable research efforts are needed to quantify PR in terms of interactive effects of PI, Ns, and Bs for the semi-arid and subhumid tropical and subtropical soils and even for other regions. The Dr is affected by acidification (excess Al, Mn), sodification (excess Na), salinisation (excess salts), toxic metals, and synthetic organic contaminants.

The effect of wind erosion on crop productivity has been even more difficult to assess than that of water erosion, especially for Oxisols, Inceptisols, and Vertisols (Zobeck and Bilbro 2001, Nordstrom and Hotta 2004; Santra et al. 2017). Lyles (1977) estimated a yield reduction of about 2 kg grain ha^{-1} mm^{-1} of soil loss. On coarse-textured Alfisols and Ultisols, especially in the semi-arid regions, reductions in crop productivity from soil loss by wind erosion are likely to be drastic (FAO 1979; Marsh 1981), although long-term crop productivity rates (PR) are not available for wind erosion. Santra et al. (2017) estimated the impact of wind erosion on crop productivity and economic loss in the arid region of western Rajasthan, India. They reported that grain yield losses of 331 and 195 kg ha^{-1} $year^{-1}$ for groundnut (peanut) and pearl millet, respectively, were affected very severely by the wind erosion category. Wind eroded economic loss was higher for groundnut and pearl millet, about 194–12465 ha^{-1}. In this study, they considered the indirect effect of soil fertility decline; however, future studies are required to assess the direct impact/damages on crops by wind erosion.

3.2 Nutrient Removal in Produce

3.2.1 Natural Ecosystem and Land Clearing

In a natural ecosystem, Ns (Equation 3.10) is a function (f) of state factors of soil formation (Jenny 1961), that is

$$Ns = f(cl, o, r, p, t) \tag{3.10}$$

where
cl = climate
o = organism/vegetation
r = relief/topography
p = parent material
t = time

In steady-state climax vegetation, change in Ns over time in the vegetation-soil system would be negligible since the rate of loss of a nutrient element would be balanced by its rate of accretion. This is expressed by Equations 3.11 and 3.12:

$$d\,Ns/dt = -rNs + aA \tag{3.11}$$

since at steady-state

$$d\,Ns/dt = 0, rNs = aA \tag{3.12}$$

where r = rate of loss
a = rate of accretion
A = amount of accretion of a nutrient element

Distribution of a nutrient element between vegetation and soil system in a natural ecosystem at steady-state generally differs from semi-arid to subhumid regions; in the semi-arid regions, a large proportion of nutrient elements is contained in soil and root biomass, whereas in the subhumid regions, it is contained in the above-ground vegetation and litter layer (Zhang et al. 2023). Thus, the appropriate method of land clearing should be used to minimise the loss of nutrient elements from such operations (Mosier et al. 2021).

3.2.2 Agroecosystem and Nutrient Depletion

Agroecosystems derived from the natural ecosystem generally result in nutrient element depletion because the rate of loss exceeds the rate of accretion ($rNs > aA$) of a nutrient element (Barrow 1991; Montgomery 2007). Besides the likely loss due to erosion and runoff (Sect. 3.1.2), nutrient element removal in the economic produce of crops consistently depletes the soil of a number of nutrients if their

Table 3.3 Nutrient element removal in the economic produce of a number of crops grown in sub-tropical and tropical regions

Crop	Estimated yield (Mg ha^{-1})	Nutrient concentration (kg Mg^{-1})							References[a]
		N	P	K	Ca	Mg	S	Zn	
Wheat	2.0	23.1	3.4	4.5	0.5	1.4	1.8	0.029	1, 3, 4, 6,13
Barley	2.0	20.0	2.7	5.0	0.4	1.2	2.0	0.019	6
Oats	2.0	21.3	3.5	4.6	0.8	1.4	3.8	0.013	1
Corn	3.0	23.8	3.3	5.0	0.2	0.9	2.8	0.072	4, 5, 7, 8, 9
Sorghum	3.0	18.8	3.7	4.8	0.3	1.1	1.6	0.040	4, 8, 9,
Rice	2.0	14.6	3.6	3.6	9.1	4.6	3.2	0.040	10, 15
Millet	1.5	26.9	7.3	15.5	1.6	0.6	2.6	0.029	4, 8, 11, 13
Alfalfa	5.0	34.5	2.7	19.6	0.6	2.1	2.1	0.038	4, 8, 9,
Chickpea	1.0	41.8	2.0	5.3	1.3	1.9	4.0	0.020	12, 15
Cowpea	1.0	38.6	3.5	13.6	0.6	2.0	1.9	0.032	2, 13
Beans	1.0	44.7	6.2	14.4	2.0	2.2	4.5	0.025	4,
Groundnut	1.5	52.0	5.1	7.1	8.8	3.3		0.024	12,14,15,17
Soyabean	1.5	72.8	5.9	15.9	4.1	3.5		0.010	11,13
Sunflower	1.0	26.0	4.1	8.0					1,6,7,13
Cotton	2.0	37.0	7.6	16.0					4,9,17
Banana	10.0	28.7	2.1	41.3					5,9
Sugarcane	75.0	19.1	3.3	23.1					6,13
									9
									9,13
									13,19

[a] 1 – Russell (1973), 2 – Campbell (1978), 3 – Lipsett and Dan (1983), 4 – Tandon and Rego (1988), 5 – Mueller-Harvey et al. (1985), 6 – French (1991), 7 – Chabalier (1985), 8 – Gigou et al. (1985), 9 – Pieri and Oliver (1988), 10 – Thomas et al. (1990), 11 – Sanchez and Benites (1987), 12 – Misra and Maheshwari (1988), 13 – Reuter and Robinson (1986), 14 – Ahlawat (1990), 15 – ICRISAT (1984), 16 – ICRISAT (1988), 17 – ICRISAT (1989), 18 – Sanchez (1987), 19 – BSES (1979); Source: Modified from Dalal et al. (1997)

rates of accretion (e.g. fixation of atmospheric N_2 by the legume plants and other N_2 fixers, weathering of soil minerals, atmospheric deposition, fertilisers, and manures) are less than their rate of removal from the soil–plant system.

The nutrient element removal in the economic production of crops (and animals) is a necessary process of nutrient element depletion; nevertheless, it leads to soil fertility depletion except at luxury consumption (Silva 2011; Culman et al. 2019). The nutrient element removal from soil by the main cereal grain crops—wheat, rice, corn (maize), sorghum, millet, and barley (mean values)—are as follows: 14–27 (22.5) kg N Mg^{-1}, 2.7–7.3 (4.1) kg P Mg^{-1}, 3.6–15.5 (7.5) kg K Mg^{-1}, 0.2–0.5 (0.4) kg Ca Mg^{-1}, 0.9–1.4 (1.2) kg Mg Mg^{-1}, 1.6–3.8 (2.1) kg S Mg^{-1}, and 0.019–0.072 (0.037) kg Zn Mg^{-1} (Table 3.3). That these values can be taken only approximate nutrient element concentrations in the grain is illustrated for N in Fig. 3.11, where grain N concentration declined from 28 kg N Mg^{-1} to 15 kg N Mg^{-1} during 50 years of cereal cropping as a result of decreasing soil total N concentrations, hence N supply to crops. Moreover, N removal in cereal grain is much higher from Vertisols than from Alfisols (Dalal and Mayer 1986a, 1986e).

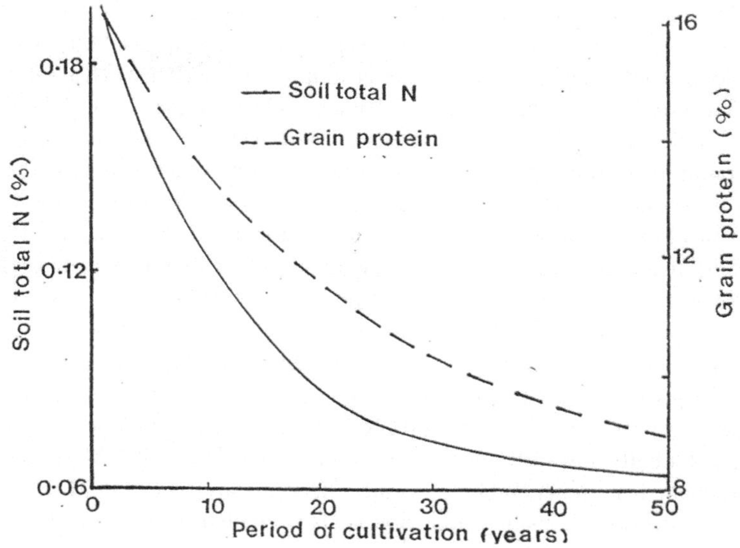

Fig. 3.11 Decline in soil nitrogen and wheat grain nitrogen with increasing periods of cultivation. Soil total N (%) = 0.068 + (0.201 − 0.068) exp (−0.086 year^{-1}). Grain N (kg Mg^{-1}) = 14.0 + (28.8 − 14.0) exp (−0.0433 year^{-1}). (Source: Dalal and Mayer 1986a, 1986e)

Table 3.4 A comparison of depletion rates of soil N by the processes of N removal in produce and soil erosion

Source	Removal in produce	Removal in soil erosion		Produce N/erosion N	
		Total	Mineralisable-N	Total N	Min. N
	kg N ha^{-1} year^{-1}				
Terrestrial lands	3.3[a]	1.46d	0.15f	2.2	21
Cultivated lands	30–40[b]	6.35e	0.64f	4.5	44
Alfisols	28.00[c]	6.35	0.32g	4.4	88
Vertisols	46.00[c]	6.35	0.64g	7.2	72

[a]Calculated from Table 3.3 mean grain-N concentration (22.5 kg t^{-1}) and total food grain production (1952.224 × 10^6 Mgt, FAO, 1992) and total land area (13.075 × 10^9 ha)
[b]Grain-N removal rate from cultivated lands (1.461 × 10^9 ha)
[c]Calculated from Dalal and Mayer (1986e)
[d]From Table 3.1
[e]Assuming 50% of sediment-N loss from cultivated lands
[f]Assuming 10% total N as potential mineralisable N (Dalal and Mayer 1987b)
[g]Calculated from Dalal and Mayer (1987b)

Since the nutrient removal from continuous cropping provides a better understanding as to how much fertility depletion takes place, it provides an estimation of how much nutrient is required to replenish to maintain fertility status for sustained crop production (Sanchez et al. 1997, Sanchez and Jama 2001, Havlin 2020), although it should be corrected for nutrient use efficiency provided that other losses are minimal.

The legume crops generally remove more N in grain from the soil–plant system (Table 3.3), although a proportion of N is derived from the atmospheric N_2 fixation (via biological nitrogen fixation), so net N removal is much less (Kakraliya et al. 2018). Excessively large removal of K in banana and sugarcane (Table 3.3) results in a rapid depletion of K fertility of soil. Moreover, N fertiliser through the incorporation of legume residue provides (additional N to succeeding cereal crops (wheat) in Vertisols). The legume residue incorporation helps to increase the yield component of wheat crops by providing an additional 18–46 kg N ha^{-1} (Regassa et al. 2023). In addition, the adoption of integrated soil fertility management (ISFM), in which organic inputs are utilised, is a potential option to reduce the complete dependence on inorganic fertiliser inputs.

A comparison of depletion rates of soil N by the processes of N removal and soil erosion shows that the former is the major process of soil fertility depletion, especially when indices of current soil productivity, that is, the ratio of produce N removal/potentially mineralisable N, are compared (Table 3.4). Moreover, since SOM-associated nutrients, such as N, are released only when SOM is oxidised/mineralised, the organic matter in the soil is also depleted (Henry et al. 2023).

3.3 Nutrient Removal in Crop Residues

While nutrient removal in the economic produce is a necessary process of soil fertility degradation, it need not be so in crop residues because the latter can be returned to the soil. In countries of developed economies, alternative uses of crop residues are minimal (Trimble and van Hook 1984; Blanco-Canqui and Lal 2009; Pasricha 2017; Regassa et al. 2023; Cooney et al. 2023), although, in those of developing economies, crop residues are used for cattle feed, building materials, fibre, and fuel for cooking and industries (Prasad et al. 1999). Crop residues when returned to the soil have a number of beneficial effects: they provide substrate for microbial and mesofauna activity which assist in nutrient cycling; increase organic matter, total N, available P, and exchangeable K—especially under no-tillage, conservation agriculture, and climate-smart agricultural practices; and affect infiltration and retention of soil water, regulate surface soil temperature, and promote aggregation (Prasad and Power 1991; Prasad et al. 1999). When crop residues are removed from the soil, not only these benefits to the soil are denied but also a substantial pool of nutrient elements is removed (Table 3.5). For example, corn leaf and stem contain higher amounts of K, Ca, and Mg than grain, and removal of these from the field leads to fertility degradation of soil (Madalcho et al. 2015).

Nutrient element concentration in different crop residues varies four-fold in N, three-fold in P, fifteen-fold in K, and five-fold in Ca but less than two-fold in Mg. Estimated mean amounts of N, P, K, Ca, and Mg contained in crop residues, respectively, are 24, 3, 41, 14, and 5 kg ha^{-1}, which were 25, 20, 65, 90, and 40% of the total N, P, K, Ca, and Mg, respectively, in the above-ground crop. Hence, the rate of soil fertility degradation would be considerably increased when crop residues are also removed along with the grains or economic produce (Malchado et al. 2015). The

Table 3.5 Nutrient element concentration in crop residues

Crop	Estimated yield ($Mg\ ha^{-1}$)	Nutrient concentration ($kg\ Mg^{-1}$)					References[a]
		N	P	K	Ca	Mg	
Wheat	3.0	5.6	0.9	9.1	1.8	1.1	1, 3
Corn	4.5	4.3	0.7	10.7	4.9	2.1	2, 4, 5, 10
Sorghum	5.0	5.9	1.2	10.1	2.9	1.5	5, 6, 10
Rice	3.5	11.5	0.8	23.2	9.5	2.3	7, 9
Millet	4.0	4.3	0.7	10.3	10.1	1.8	5, 10
Chickpea	1.5	5.7	1.2	2.3			8
Cowpea	1.5	18.6	1.3	34.7			7
Beans	2.0	15.4	1.4	19.1			1
Groundnut	1.0	12.6	0.5	10.6			5
Soyabean	3.5	10.7	1.3	15.8			2,5,10
Cotton	3.0	9.0	2.0	12.9			5,10
Mean	3.0	8.2	1.1	13.8	4.9	1.7	

[a]1 – Russell (1973), 2 – Mueller-Harvey et al. (1985), 3 – French (1991), 4 – Chabalier (1985), 5 – Pieri and Oliver (1988), 6 – Thomas et al. (1990), 7 – Sanchez and Benites (1987), 8 – Misra and Maheshwari (1988), 9 – Reuter and Robinson (1986), 10 – Power and Legg (1978); Source: Modified from Dalal and Probert (1997)

soil fertility degradation by the process of crop residue removal is especially critical for Ca in soils with low exchangeable Ca and low-activity clays such as Oxisols and Ultisols and to a certain extent acid Alfisols, especially in subhumid regions, where removal of crop residues would drastically reduce soil productivity (Sanchez and Benites 1987) in the absence of liming (Pushparajah and Buchik 1987). For example, Sanchez and Benites (1987) reported that the retention of rice and cowpea crop residues returned to an Ultisol contributed 94% Ca, 89% K, and 69% Mg accumulated by the crops. Thus, the removal of crop residues as a process of soil fertility depletion becomes significant, especially in low fertiliser input agroecosystems in the subhumid and semi-arid tropical and subtropical regions (Wani et al. 2007; Odendo et al. 2010; Malchado et al. 2015).

3.4 Oxidation, Reduction, and Leaching

3.4.1 Oxidation

The primary oxidation process that sustains biological activity in soil is the oxidation of SOM, assisting thereby the cycling of nutrient elements contained in, associated with, or complexed with organic matter (Bot and Benites 2005; Tan 2010). However, when the oxidation rate of SOM exceeds its rate of addition, organic matter depletion and hence soil fertility degradation occurs (West and Post 2002; Henry et al. 2023; Hoffland et al. 2022). This is described in detail in Sect. 4.1.

3.4.1.1 Burning

Another significant oxidation process is the burning of surface and standing biomass. Intensive and frequent burning in semi-arid and subhumid environments causes loss

Fig. 3.12 Clearing crop (wheat) residue by burning is a common practice in Central India and other developing countries to quickly clear the fields for sowing or planting the succeeding crop. (Photo: Jayaraman and Dalal)

in the gaseous form of most organic C, N, and S in the plant biomass (Young 1989; Andriesse 1989; Bhuvaneshwari et al. 2019). Furthermore, substantial losses of P, Ca, K, and Mg can occur in particulate form during a hot burn (Walker et al. 1986; Prasad et al. 1999; Jayaraman et al. 2020) (Fig. 3.12). Mandal et al. (2004) reported that burning of crop residue, particularly rice and wheat straw, resulted in the loss of essential nutrients, namely, 80% of N, 25% of P, 21% of K, and 4–60% of S. Approximately 25% of N and P, 50% of S, and 75% of K are taken up by cereal crops when these nutrients were retained in crop residues, and these nutrients are made available when returned to soil or recycled through residue management (Jayaraman et al. 2020)

Nitrogen losses from plant biomass burnt at temperatures exceeding 200 °C increase rapidly (Fig. 3.13), and up to 90% of total N in plant biomass can be lost at >350 °C (DeBano and Conrad 1978; Andriesse and Schedhaas 1987). Almost half of this N loss occurs as molecular nitrogen (Kuhlbusch et al. 1991). It has been estimated that the losses of N from plant biomass burning are 10–-50 × 10^9 kg N $year^{-1}$, about 90% of which originate from the tropics and subtropics. These losses of N are equivalent to 5–50% of terrestrial N_2 fixation and thus become particularly critical for tropical agroecosystems (Kuhlbusch et al. 1991).

Beneficial effects of the 'slash-and-burn' system of shifting cultivation, albeit temporary (1–4 years), have been observed on strongly leached Ultisols, Oxisols, and acidic Alfisols, primarily through the increase in soil pH (Kyuma et al. 1985; Andriesse 1989), decrease in exchangeable Al (Serrão et al. 1979), enhanced nitrification rates (Montagnon and Buschbacher 1989), and increase in available P (Serrão et al. 1979; Lopes et al. 1987) and exchangeable cations (Andriesse 1989) but loss of biodiversity (Hauser and Norgrove 2013). However, plant biomass burning, normally encountered in the field, has only a small effect on the properties of neutral and calcareous Vertisols except for inorganic P (Kwari and Batey 1991), but it can be used as a management option for disease control. In the long-term, however, even on Ultisols and Oxisols, agricultural productivity declines, due to loss of cations by erosion, produce removal and, through intensive leaching and reduced cation

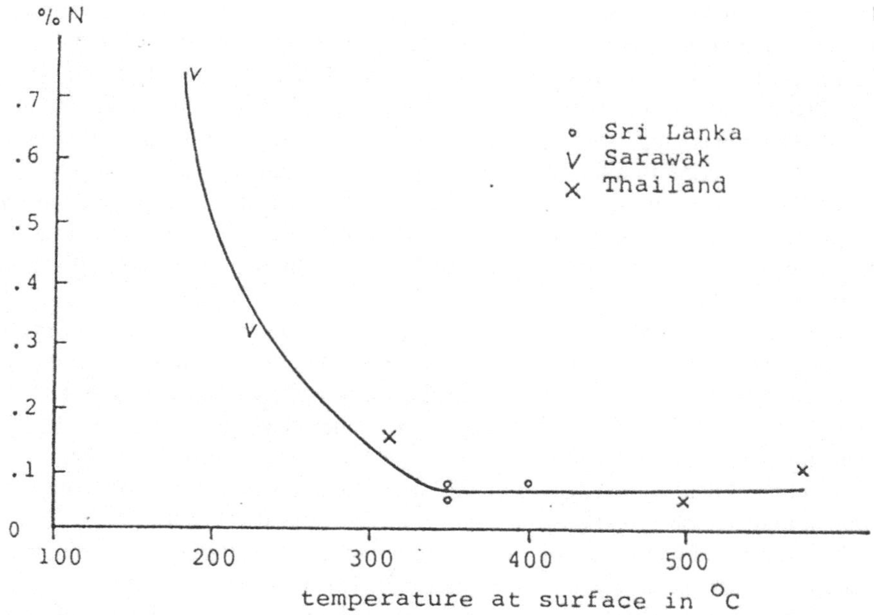

Fig. 3.13 Effect of burning temperature of plant biomass and N content of ash. (Source: Andriesse and Schedhaas 1987)

exchange capacity, decline in organic matter, increased soil acidity (Lal 1989; Juo and Manu 1996; Das et al. 2020), and release of CO_2 due to forest clearing, which contribute to greenhouse gas (GHG) emissions and hence climate change (Tinker et al. 1996). However, in the process of 'slash-and-burn' farming, a larger portion of nutrients (K, Mg, Ca, nitrate, and sulfate) are also lost through runoff/erosion or leaching (Grimme and Juo 1985; Juo and Manu 1996).

3.4.1.2 Sulfide Oxidation

Chemical and bacterial (*Thiobacillus ferrooxidans*) oxidation of pyrite (FeS_2), when pyritic soils are drained, produces sulphate with a concomitant production of sulphuric acid and toxic concentrations of Fe, Al, and Mn and high salinity, resulting in relatively infertile acid sulphate soils (pH <4) (Bloomfleld and Coulter 1974; van Breemen 1982; Charoenchamratcheep et al. 1987; Ward et al. 2004; Sarangi et al. 2022), and these soils pose serious problem to plant, other biota, as well as environmental risk (Dent and Pons 1995). The oxidation of both Fe^{2+} and S^{-} occurs as follows (Kelley and Tuovinen 1988; Tan 2010):

$$2FeS_2 + 1.5\,O_2 + H_2O \rightarrow 2Fe^{3+} + 4SO^{-} + 2H^{+} \tag{3.13}$$

or

$$2FeS_2 + 2H_2O + 7O_2 \rightarrow 2FeSO_4 + 2H_2SO_4$$

The acid sulphate soils (ASS) or cat-clays occur in the recent coastal plains in the tropics and subtropics (Kawalec 1973; Sanchez 1976; Dent and Pons 1995; Isbell 2016). These soils are classified as Sulfaquepts and Sulfaquents (Brady 1996). They are generally used for paddy rice production because, under waterlogged conditions, pyrite oxidation is suppressed. By removing H_2SO_4 by lateral surface drainage, these soils have also been used for pineapple production in Thailand (van Breemen 1975, 1976). In Guyana, the acid sulphate soils are kept in paddy rice production by occasional treatment with seawater to remove acidity. In India, ASS occurs in the coastal and lower Ganga River system, and these lands are used for paddy cultivation (Sarangi et al. 2022).

The oxidation of previously submerged paddy rice soils (reduced soils) generally results in phosphate sorption by poorly crystalline ferric hydrous oxides or re-oxidised iron [(Fe^{+2} + $0.25O_2$ + organic C → Fe^{3+} + organic C + e^-; Fe^{3+-} + H_2PO_4 + H_2O + $0.5O_2$ → $FePO_4$. $2H_2O$) Willett and Higgins 1978], thereby reducing phosphate availability to the succeeding arable crops. Therefore, it is recommended that in the rotation of rice with arable crops, fertiliser P should be applied to the arable crop (Ru-Kun et al. 1982; Sanyal and De Datta 1991).

It is suggested that oxidation of Cr^{3+} to chromate anion (CrO_4^{-2}) in soils may lead to increased metal toxicity (McBride 1989). This is because the chromate anion is generally more soluble than Cr^{3+} in soil and much more toxic to animals and other biota. However, there is a scarcity of information available to evaluate the extent and magnitude of chromate concentration in soils under different oxidation states.

3.4.2 Reduction

Besides increasing the solubility of some metal nutrient elements occasionally to toxic concentrations (Fe, Mn), the reduction process is involved in soil fertility degradation primarily through anaerobiosis (reviewed by Fausey and Lal 1990) and reduction of nitrates in soil to gaseous products, NO_x, and N_2 (denitrification) (Schepers and Raun 2008).

The sequence of nitrate reduction is represented as follows: NO_3^- → NO_2^- → NO → N_2O → N_2 (2NO + $12H^+$ + $2e^-$ → N_2 + $8H_2O$). This is carried out mostly by facultative microorganisms who utilise oxygen of NO_2 (nitrite), NO (nitric oxide), and NO_x (nitric oxide + nitrite) as a hydrogen acceptor only in the absence of free oxygen (Knowles 1982, 2005). Firestone (1982) listed four requirements for denitrification: (i) the presence of bacteria (archaea, fungi) possessing the metabolic capacity, (ii) suitable electron donors such as organic C compounds, (iii) reduced S compounds, or (iv) molecular hydrogen, H_2, anaerobic conditions or restricted O_2 availability, and NO_3, NO_2, NO, or N_2O as terminal electron acceptors (Schepers and Raun 2008).

Generally, the loss of soil nitrates through denitrification increases with an increase in easily decomposable organic matter, soil temperature, and nitrate concentration and a decrease in oxygen supply to 4–17% (anaerobic hot spots) (Dalal et al. 2003). Greenwood (1975) concluded that O_2-free zones (anaerobiosis) are

Table 3.6 Denitrification rates in a tropical and a temperate forest

Successional stage	Treatment	Denitrification rate (g ha^{-1} h^{-1})	
		Tropical	Temperate
Early	Unamended	2.43	110.0
Mid	Unamended	0.48	1.5
Late	Unamended	1.50	3.6
	Water	11.07	0.6
	Water + nitrate	9.16	106.8
	Water + glucose	106.4	

Source: Groffman et al. (1988)

likely to occur in soils having <10% air-filled pore space and water-saturated aggregates >9 mm in radius. However, anaerobic microsites in the soil where denitrification occurs are the sites of intense respiratory activity (hot spots) rather than passive anaerobiosis (Craswell 1978; Knowles 1982; Schepers and Raun 2008). The situation is complicated by the fact that the requirements for active denitrification (available organic C, nitrate, denitrifying organisms, and oxygen-depleted zone) may be randomly distributed in the soil matrix but must all coincide for denitrification to occur (Dalal et al. 2003). Thus, it varies both in space and time, and therefore, it is extremely difficult to predict the overall denitrification rate for a given soil and environment (Schepers and Raun 2008).

A comparison of potential denitrification rates in a lowland tropical forest and a temperate forest shows that in both ecosystems, denitrification is relatively high in early and late successional communities and that the absence of easily decomposable C and NO_3-N limits the denitrification rates in a mid-successional forest ecosystem to a low level (Table 3.6) (Groffman et al. 1988; Vitousek and Matson 1988), but not negligible in tropical rainforest ecosystem (Dalal and Allen 2008). However, when a forest ecosystem or permanent grasslands are converted into agroecosystems, the relative synchrony between the supply and demand of NO_3-N is broken, thus increasing the frequency of high NO_3-N concentrations in soil. Furthermore, the capacity to use water by plants is diminished and becomes discontinuous (clean fallow), so the frequency of waterlogging also increases. Along with deteriorating soil structure (aggregate breakdown) and increased compaction, oxygen supply is likely to be reduced (Hillel 2008; Lal 1998). Provided that easily decomposable organic C is available, the denitrification rates in soils under agroecosystems are likely to be much larger than those under natural ecosystems. Also, the N losses through denitrification are accentuated by increasing use of N fertilisers where the ratio of N_2O/N_2 increases considerably (Mosier et al. 1991), especially in acidic soil (Wang et al. 2016), which is further exacerbated depending on the forms of fertiliser application (Mulvaney et al. 1997). As a result, as much as 30% of added N can be lost as gaseous N (Hauck 1984). However, N fertiliser formulation with a nitrification inhibitor (3,4-dimethylpyrazole phosphate) markedly decreases N_2O emissions from acid soil (Dalal et al. 2003; Das et al. 2022). Nitrous oxide emissions were also reduced by lime application, but the application of both nitrification inhibitor–

formulated N fertiliser and liming decreased nitrous oxide emissions by 90% in an acidic soil used for sugarcane cropping (Das et al. 2022).

Measured losses of added N, presumably through denitrification, vary considerably, from nil to 30% in Vertisols (Moraghan et al. 1984; Craswell and Strong 1976; Freney et al. 1985) to as high as 50% in Alfisols (Gigou and Dubernard 1979). For example, on a Vertisol, N losses, mainly through denitrification, increased from 6% with 800 mm of rainfall to 26% with 1200 mm of annual rainfall (Burford et al. 1989).

Buresh and De Datta (1990) summarised the results of direct measurement and indirect estimation of denitrification losses from paddy rice fields in the Philippines, Thailand, and Indonesia. The gaseous N losses recovered as (N_2 + N_2O)-N were much lower (0.1–2.2%) than that estimated by the N balance (10–56%) of the applied urea (^{15}N). The latter presumably also included N loss by NH_3 volatilisation. Denitrification was limited by the nitrate supply rather than by available C. Thus, nitrification–denitrification occurs under alternate soil drying and wetting conditions which favour NO_3 formation both during the rice cropping period and between rice crops.

Measurements of denitrification in natural ecosystems as well as agroecosystems in the absence of fertiliser N additions are limited by suitable 'in situ' techniques. Estimates of N losses by denitrification process, therefore, vary widely, from 0 to 15×10^9 kg N $year^{-1}$ (Knowles 1982; Ryden and Rolston 1983). Further improvements in the methodology using ^{15}N-natural and ^{15}N-tracer mass abundance and a better understanding of denitrification processes may provide a better estimate of denitrification losses in different ecosystems (Schepers and Raun 2008).

3.4.3 Leaching

Leaching is often the most important process of nutrient loss in agroecosystems other than that accounted for by the removal of produce (Sect. 3.2). It is also responsible for the acidification of Ultisols, Oxisols, and acid Alfisols, especially in subhumid regions, with the leaching losses of basic cations and basic cations removed in the produce (Sect. 3.5).

The depth of leaching of a nutrient such as NO_3^{-1}-N is determined, at least in part, by the amount of water passing through the soil, that is

$$Q_w = W_p - W_r - W_{ev} - W_{tr} - \_W \tag{3.14}$$

where Q_w = amount of through-flow water
W_p = precipitation (and irrigation)
W_r = runoff
W_{ev} = evaporation from soil
W_{tr} = transpiration from crops
_W = increase in soil water content

A number of components, therefore, affect the amount of water available for leaching. For the one-dimensional transport of a solute through the soil, the convection-dispersion equation, with some modifications, has been widely used. The partial differential equation for the one-dimensional solute transport is

$$C/t = D^2C/x^2 - v.C/x - P/q.S/t \quad (3.15)$$

where C = solution concentration (ML^{-3})
t = time (T)
x = distance of displacement (L)
D = dispersion coefficient (L^2T^{-1})
v = interstitial flow velocity (LT^{-1})
S = amount of solute sorbed (MM^{-1})
P = soil bulk density (ML^{-3})
q = volumetric water content (L^3L^{-3})

The highly variable nature of the dispersion coefficient and interstitial flow (pore water) velocity limits the applicability of this convection-dispersion equation (Eq. 3.15) or other deterministic models (White 1988). Using stochastic models, Amoozegar-fard et al. (1982) showed that the variability in interstitial flow velocity had a greater effect on solute transport than the dispersion coefficient. Since the nutrients are not uniformly distributed in the soil matrix (due to sorption for immobile nutrients, sites of biological transformation, plant uptake) only a fraction of Q_w (Eq. 3.14) may actually be involved in the solute transport, that is, the transport volume, V_{st}, may differ considerably both in time and space. For NO_3-N leaching, White (1988) derived the following relationship:

$$N = 10^{-3}\ C_o\ V_{st}(1 - \exp\ (-10d_o t/V_{st}) \quad (3.16)$$

where N = quantity of NO_3-N leached in kg ha^{-1}
C_o = mean concentration of NO_3-N in the soil solution initially
d_o = mean leaching rate (mm h^{-1}) which can be approximately estimated from Eq. 3.14
t = time in hr
V_{st} = transport volume (L L^{-3})

The use of both deterministic and stochastic transfer models may describe solute leaching satisfactorily in some field situations. In the light of very large temporal-spatial heterogeneity of some nutrients in the soil, even empirical approaches may be sufficient for making management decisions (Pal and Poonia 1990).

Of all the often-limiting nutrients for crop growth, N as NO_3-N is readily leached from the soil below the root zone (Nakamura et al. 2004). Ritter (1989) reported that leaching losses of N as NO_3 can be large (4–350 kg N ha^{-1} $year^{-1}$) when soil NO_3-N concentration is high and water movement is large. Thus, both water movement and nitrogen management are important in controlling nitrate leaching.

Table 3.7 Nitrate leaching below 1 m depth in semi-arid Vertisols in relation to the period of cultivation for cereal cropping

Soil depth	NO_3-N kg ha^{-1} after cultivation period (year)					
	0[a]	0.5[a]	12[a]	16[b]	19[c]	20[a]
0–1 m	28	84	134	36	95 (400)	95
>1 m	1	30	147	260	223 (635)	196
Sampling depth (m)	3	3	3	2.8	2.5	3
Annual rainfall (mm)	610	610	610	750	716	610
Vegetation/crop	Forest	Wheat	Wheat	Sorghum	Sorghum	Wheat

Source: [a]R.C. Dalal, Unpublished data; [b]Standley et al. (1990); [c]V.R. Catchpoole, personal communication. In parentheses, NO_3-N contents when 100 kg N ha^{-1} year^{-1} of fertiliser applied to sorghum

Table 3.8 Nutrient losses by leaching below the root zones of peanut, cotton, and maize in Alfisols

Location	Rainfall (mm)	Crop	Water drained (mm)	Nutrients leached (kg ha^{-1} year^{-1})			
				N	Ca	Mg	K
Bambey, Senegal	507	Peanut	101	25	39	8	4
Mauoua, Cameron	683	Cotton	83	2	31	7	1
Bouake, Ivory Coast	532	Cotton	260	7	13	4	2
	633	Corn	210	6	26	16	2

Source: Adapted from Pieri (1987), ISBRAM, Bangkok, Thailand

Thus, in semi-arid and subhumid regions, with rainfall of up to 1200 mm a year, leaching losses from Oxisols, Ultisols, and Alfisols growing crops were only 8—15 kg N ha^{-1} year^{-1} (Pieri 1979; Vachaud et al. 1982; Chabalier 1985), whereas more than 80% of ^{15}N-labelled NO_3-N was lost within 2 years from an Ultisol with an annual rainfall of 2400 mm (Wong et al. 1987). In the long term, however, even from relatively less permeable soils, Vertisols, substantial leaching losses of NO_3-N do occur, especially when the soil is cleared of native vegetation and/or fertiliser N applied (Table 3.7). Moreover, soil types and amount of N application play a significant role in the leaching of NO_3-N. Wang et al. (2019a, b) estimated NO_3 leaching from the global dataset (i.e. 324 sites) and inferred that soil nitrate leaching emission factor (EF) may not remain constant, but it increases rapidly with increasing rates of N application. They adopted a methodology, i.e. EF response to increasing N input (ΔEF) was found more accurate and sensible in estimating the leaching loss of NO_3-N.

Besides the leaching losses of NO_3-N, basic exchangeable cations, such as Ca^{2+}, Mg^{2+}, and K^+ are also leached substantially from Oxisols, Ultisols, and acid Alfisols (Table 3.8) so that the major cause of fertility depletion in these soils is generally the deficiency of basic exchangeable cations. Exchangeable cations are retained in the soil on its net negative charge surfaces (due to permanent and variable charges). On a variable charge surface where protons are the potential-determining ions, the net surface charge can be positive, zero, or negative, depending upon the difference between the actual soil pH and the pH where the net surface charge is zero (pH_o), that is

$$s_o = (2nekT/p)^{0.5} \ \sinh \ 1.15_z \ (pH_o - pH) \quad (3.17)$$

where s = surface charge
n = concentration of electrolyte solution of dielectric constant 'e'
z = counterion valency
k = Boltzmann constant
T = absolute temperature

Thus, the magnitude of the surface charge is governed by the absolute difference between pH and pH_o and by the concentration and composition of the equilibrium electrolyte solution. In soils (Oxisols, Ultisols, and acidic Alfisols high in iron and aluminium oxides) that derive most of their cation retention capacity from the SOM, loss of SOM results in an increase in pH_o in these soils (Gillman 1984), according to the following relationship:

$$pH_o = 3.4 + [(7.3 - 3.4) \ \exp \ (-0.52 \text{ organic C } (\%)], r^2 = 0.85 \quad (3.18)$$

According to this relationship, the decline in organic C from 4% to 1% results in an increase in pH_o from 3.9 to 5.7. Assuming the pH of the soil is 5.5, the loss in organic C from 4% to 1% virtually wipes out its capacity to retain basic cations. Gillman (1984) observed that an Oxisol under a native forest that had been under cultivation for sugarcane for 50 years had 4.4% and 1.8% organic C contents, respectively, had pH_o of 3.8 and 4.7, and had net negative charges of 3.5 and 0.7 cmol kg^{-1}, respectively. Consequently, the exchangeable Ca^{2+}, Mg^{2+}, and K^+ declined from 2.0 to 0.3, 1.2 to 0.1, and 0.20 to 0.08 cmol kg^{-1}, respectively. Liming or fertilisation alone is only a short-term solution for fertility restoration of these soils once the organic matter is depleted because of their poor cation retention capacity (Lungu 1987), and thus, such soils require better management strategies to conserve organic matter (Lopes and Cox 1977).

3.5 Acidification, Sodification, and Excessive Element Load

3.5.1 Acidification

Soil acidity is one of the major problems in the agricultural production system as it deteriorates soil health and significantly reduces crop productivity (Brady 1996; Butterly et al. 2022). Soil acidification is a result of an irreversible flux of protons to the soil. According to Van Breeman et al. (1984), proton sources include (a) atmospheric inputs of acid or potentially acidic substances, such as SO_2; (b) net assimilation of cations, such as Ca^{2+}, Mg^{2+}, K^+, and Na^+, by vegetation; (c) mineralisation of anions from organic matter; (d) deprotonation of week acids;

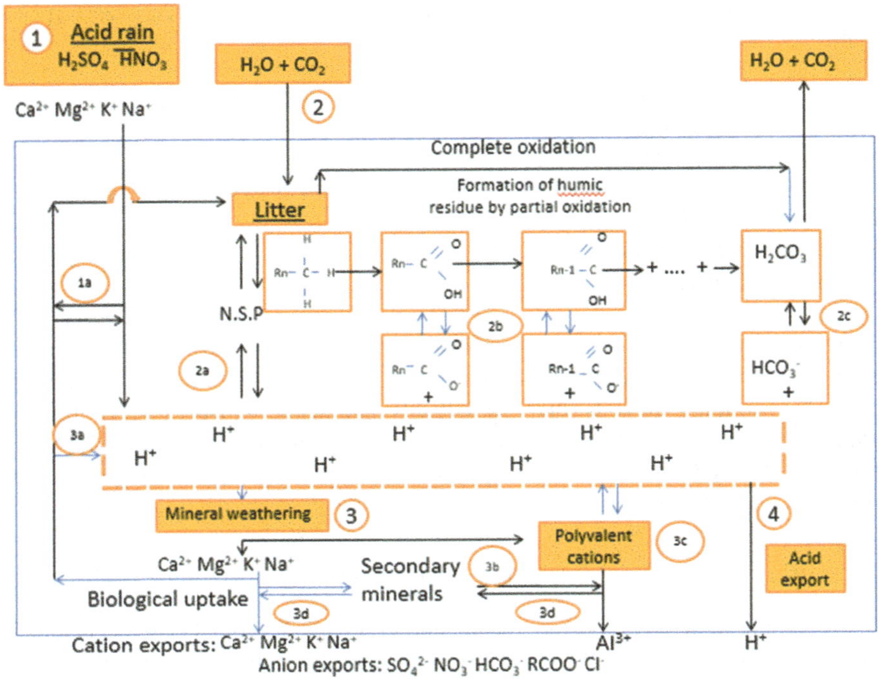

Fig. 3.14 Major sources and sinks of soil acidity in soil. (Modified from Krug and Frinck 1983; Dalal and Probert 1997)

(e) oxidation reactions, such as the formation of acid sulfate soils (Sect. 3.4) and H_2CO_3; (f) precipitation of cations; and (g) mineral weathering of anionic components.

The major sources, reactions, and sinks of protons in soil are summarised in Fig. 3.14. Rain (1) is a source of acidity; Lesack and Melack (1991) reported a mean pH of 4.9 for the rainfall in the central Amazon Basin. However, rainfall composition may be altered before reaching the soil (1a). Biological processes (2) are the net sources of acidity, but they also serve as a substantial sink in acid soils through the production of weak organic acids, such as H_2CO_3 (2b), which are ultimately converted to CO_2 and H_2O (2c). Mineral acids (2a) can be cycled rather tightly through plant and microbial biomass, with some S and N lost to the atmosphere and some S and P converted to solid phases. Weathering of minerals (3) generally consumes acid in excess of cation export (3d), although in relatively weathered soils such as Ultisols and Oxisols, the latter exceeds the proton consumption by soil. Secondary minerals (3b) and hydrolysis products of Al, Fe, and Mn (3c) accumulate in soil. Net cation uptake by vegetation (3a) can contribute to acidification. If rainfall is less acid than the soil solution, it can be acidified by metal hydrolysis (3c). Removal of cations (Ca^{2+}, Mg^{2+}, K^+, Na^+, and Al^{3+}) (3d) and protons (4) is accompanied by the removal of anions, such as SO_4^{2-}, NO_3^-, HCO_3^-, $RCOO^-$, and Cl^- (Krug and Frinck 1983).

Acid rain (1) is a source of acidity, but the composition may vary before it enters the soil (1a). (2). Biological processes are net sources of acidity; (2a) mineral acid can be recycled with some portion of S and N lost to the atmosphere and some S and P converted to solid phases, (2b) production of weak organic acid, and (2c) conversion of CO_2 and H_2O. (3). Weathering of minerals generally consumes acid in excess of cation export (3d), as secondary mineral (3b) and hydrolysis products of Al, Fe, and Mn accumulate in soil (3c). Aggrading vegetation results in net cation uptake (3a) which can contribute to acidifications. Sometimes, rain consists of less acid than soil solution, which can also be acidified by hydrolysis (3c). Cation (2d) and acid (4) export is charge-balanced principally by anions shown at the bottom.

Rowell and Wild (1985) prepared a proton budget to indicate a comparative contribution to soil acidity from some of the sources identified in Fig. 3.14. Contributions of organic acids in acid soils and from vegetation vary from 0.1 to 0.7 kg H^+ ha^{-1} $year^{-1}$ and that from cation excess in vegetation and NH_4 oxidation and leaching vary from 0.5 to 2 and 4 to 6 kg H^+ ha^{-1} $year^{-1}$, respectively; oxidation of N, P, and S from SOM and leaching after the clearance of vegetation contributes up to 10 kg H^+ ha^{-1} $year^{-1}$ (50 kg $CaCO_3$ ha^{-1} $year^{-1}$ is required to neutralise the acidity of kg^{-1} H^+) to the total acidity in the soil. It is the latter source of acidity in soil that is primarily considered in this section. Removal of biomass and agricultural produce further removes basic cations and thus contributes to soil acidification, particularly for acidic soil.

In subhumid and, to a lesser extent, semi-arid tropics and subtropical regions, with intense weathering and leaching, most Oxisols, Ultisols, and acidic Alfisols are depleted of many of their exchangeable basic cations; *thus, H^+ and exchangeable Al^{3+} (and Mn) become the dominant ions on the soil exchange sites and the soil solution, resulting in soil fertility degradation through soil acidification*. The problem is accentuated when these soils are cleared of their native vegetation for continuous cultivation and cropping. The rapid decline of organic matter in these soils under cropping is generally accompanied by the rapid decline in soil pH, for example, from 6.7 to 3.7 in 15 years in an Ultisol (Fig. 3.15), increase in Al saturation of the soil exchange sites (Obi 1989), increase in soil pH_o, decrease in effective cation exchange capacity (ECEC), and hence reduced retention and increased leaching of Ca^{2+}, Mg^{2+}, K^+, and Na^+ (Kang and Juo 1986) in soil (Sect. 3.4.3). Consequently, the crop production declines rapidly (Lungu 1987).

Since soil pH is closely related to soil solution and exchangeable Al, Bouldin et al. (1987) observed that with the increase in Al saturation of the exchange sites, the relative grain yield of maize also declined; the reduction in yield to 50% occurred at 35% Al saturation in an Oxisol from Puerto Rico, about 68% and 74% Al saturation in an Oxisol from Brazil and an Alfisol from Nigeria, respectively (Fig. 3.17).

It is obvious that percentage Al saturation, therefore, does not provide a precise measurement of soil fertility degradation due to differential increase in acidity in different soils. The activity of Al^{3+} in soil solution has been proposed to be a better index of Al toxicity to plants (Pavan et al. 1982; Bruce et al. 1989; Zhao et al. 2022).

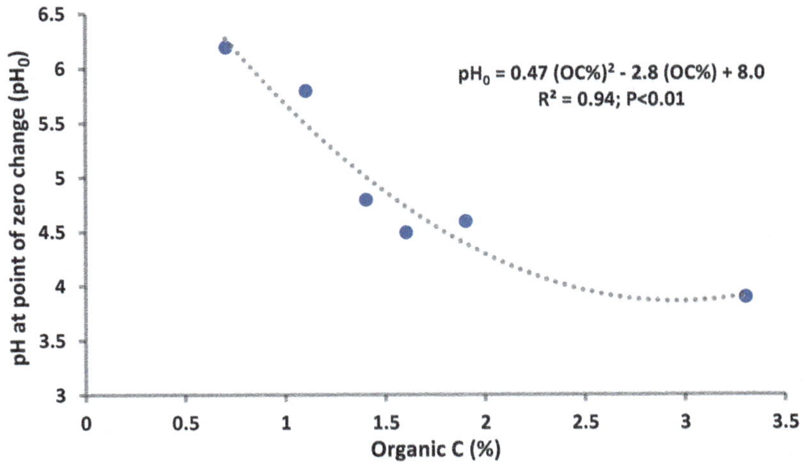

Fig. 3.15 Decrease in soil organic C results in the increase of pH at the point of zero charge (pH_0) in Oxisols. (Source: redrawn and developed by the authors from data from Gillman 1984)

Table 3.9 Decrease in base saturation as a soil acidification index

	Decrease in base saturation (% $year^{-1}$)a	
	Initial base saturation (%)	
Extent of soil degradation	<50	>50
None to slight	1.25	2.5
Moderate	1.25–2.5	2.5–5
High	2.5–5	5–10
Very high	>5	>10

Source: FAO (1979). a Base saturation (%) = 100× (Ca^{2+} + Mg^{2+} + K+ + Na+)/CEC) in 0–0.3 m soil layer

It can be calculated from the percentage Al saturation (Al sat %) and ionic strength of the soil suspension of 1:5 soil:water (I_{ss}, mM) from the following relationships:

$$\mathrm{Ln}\left(\mathrm{Al}^{3+}\right) = -6.97 + 1.96\ \ln\ I_{ss} + 0.078\ \mathrm{Al\ sat}\%, r^2 = 0.947 \tag{3.19}$$

where Al^{3+} = Al^{3+} activity in soil solution (mM)

However, in the presence of organically complexed Al, such as in Oxisols and Ultisols containing a relatively high amount of organic matter, even Al^{3+} activity is not a reliable index of soil fertility degradation due to soil acidity, probably because of the following reasons.

(i) It cannot be accurately determined in such systems.
(ii) Other Al ionic species may be important.
(iii) The dissociation constants of Al-organic complexes vary widely in different soils.
(iv) Crop species and cultivars vary widely in their tolerance to a given Al concentration in soil solution.

FAO (1979) has provided another parameter to assess soil fertility degradation due to soil acidity. This is based upon the rate of decrease in the base saturation of the exchange sites in soil (Table 3.9), primarily due to leaching losses (Takala 2019). Factors that affect this parameter are climate (precipitation > potential evapotranspiration + soil water storage), soils (dominantly kaolinitic and sesquioxidic soils, especially with reduced organic matter concentration), topography (relatively level), and anthropogenic factors (removal of natural vegetation, deforestation, acid rain, the use of acidifying fertilisers and crop species, the removal of basic cations in produce, and excessive irrigation).

Soil acidification following certain legume pastures does occur. In Ultisols of semi-arid regions of Australia, subterranean clover grown with superphosphate additions has resulted in a decrease in soil pH from 6.0 to about 5 within 50–55 years and with a substantial increase in Al and Mn in soil solution, progressively extending down to 0.6 m depth (Williams 1980; Bromfield et al. 1983; Sumner et al. 2003). Scott et al. (2000) reported that severe acidity adversely affects the tolerant pasture species in the long term in central and southern New South Wales and north-eastern Victoria, Australia.

Development and persistence of sub-soil acidity poses a serious problem for crop production (Scott et al. 2000; Dang et al. 2006). Haynes (1982, 1984) and Shainberg et al. (1989) have discussed in detail this process of fertility degradation (Haynes and Williams 1993).

Another consequence of soil acidification is the increase in phosphorus fixing capacity, especially in soils high in oxides of Fe and Al contents (Sanchez and Uehara 1980; Brady 1996). The amorphous forms of these oxides convert available P into unavailable forms probably through several mechanisms: formation of multidentate bonding with P, then rearrangement of -OH surfaces of Fe and Al to chemisorb P, and finally formation of distinct phases of Fe-P and Al-P in soil (Sanyal and de Datta 1991).

Soil acidification as a process of fertility degradation primarily involves one or more of the following: soil pH (<5); excessive active Al^{3+} (>4 mM); Al saturation (>60%); base saturation (<50%); deficient concentrations of Ca^{2+}, Mg^{2+}, and K^{+}; loss in effective CEC (pH dependent charge and loss in organic matter); high P fixation; low microbial activity and diversity; and the interactive effects of these on vegetation (plant species and plant cultivars) (Fig. 3.16). Soil and crop management practices must consider these aspects to arrest or reverse the soil acidification process of fertility degradation in many Oxisols, Ultisols, and acidic Alfisols.

3.5.2 Sodification

Sodification is the process whereby Na^{+} builds up in preference to other cations such as Ca^{2+}, Mg^{2+} (M^{2+}), and K^{+}, on the cation exchange complex of the soil colloids, that is

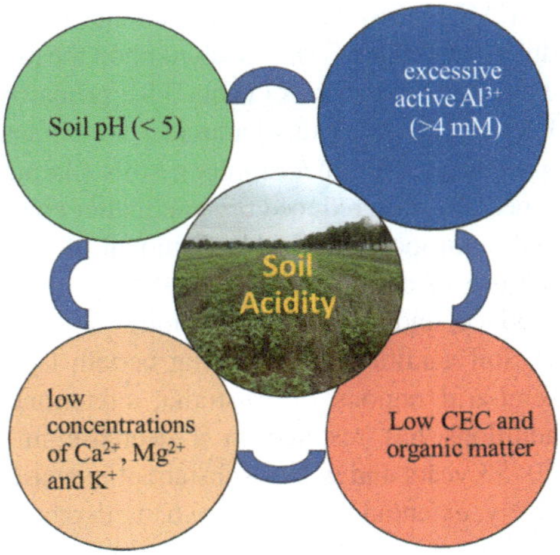

Fig. 3.16 Soil acidification process in soil. (Source: Dalal and Jayaraman)

$$M^{2+}\ \text{clay} + 2Na^{2+} \rightarrow Na - \text{clay} + M^{2+} \tag{3.20}$$

$$M^{2+} + 2HCO_3 \rightarrow MCO_3 + CO_2 + H_2O \tag{3.21}$$

$$M^{2+} + 3H_4SiO_4 + 2HCO_3 \rightarrow M\ Si_3O_6\ (OH_2) + 6H_2O + 2CO_2 \tag{3.22}$$

The divalent cations, Ca^{2+} and Mg^{2+}, precipitate out as carbonates (calcite and dolomite) and silicates such as sepiolite $Mg\ Si_3O_6\ (OH)_2$ in the process (Eqs. 3.21 and 3.22) and thus result in the increasing concentrations of Na^{+} (and K^{+}) ions in the soil solution and on the exchangeable sites (Brady 1996; Rengasamy et al. 2022). This is also accompanied by the increase in soil pH and residual alkalinity ($[HCO^{-}] + 2[CO^{-}] + [OH^{-}] - [H^{+}] - 2[Ca^{2+} + Mg^{2+}]$) since the concentration of divalent cations, Mg^{2+} and Ca^{2+}, in soil solution generally declines and sodium adsorption ratio SAR ($SAR = Na^{+}/(Ca^{2+} + Mg^{2+})^{0.5}$ in mmol L^{-1}) increases in the process of sodification (VanBeek and VanBreeman 1973), especially in semi-arid and subhumid regions.

3.5.2.1 Surface Soil Sodicity

Sodium naturally occurs in soil from mineral weathering, marine sources, rainfall, and/or aeolian dust. Under stable climatic and vegetation systems, the hydrologic cycle (rainfall, rate, and amount of water movement through the soil profile evaporation and drainage) controls the amount of salt retained in the soil profile. Land clearing, cropping, and irrigation affect the hydrologic cycle in soil and thus change the salt balance and Na in soil. Depending upon the landscape, a portion of the landscape may be (i) a recharge area, (ii) a transmission area, or (iii) a discharge area for water and hence for salt movement (Shaw et al., 1988, 1994). Thus, the rate and extent of sodification depend not only on the amount of Na present in the soil but also on the rate of accretion from the adjoining landscape or irrigation waters.

Table 3.10 Increase in exchangeable sodium percentage (ESP) with continuous cultivation and cereal cropping

Soil series	Clay (%)	Period of cultivation (years)	Increase in ESP		
			%[c]	% $year^{-1}$	% $year^{-1}$ 100 g^{-1} clay
Waco[a]	72	26.2	35	1.3	1.9
Thallon[b]	59	12.3	27	2.2	3.7
Langlands-Logie[b]	49	18.0	36	2.0	4.1
Cecilvale[b]	40	18.2	42	2.3	5.8

[a]Typic Pellusterts, [b]Typic Chromuserts, [c] From Dalal and Mayer (1986a)

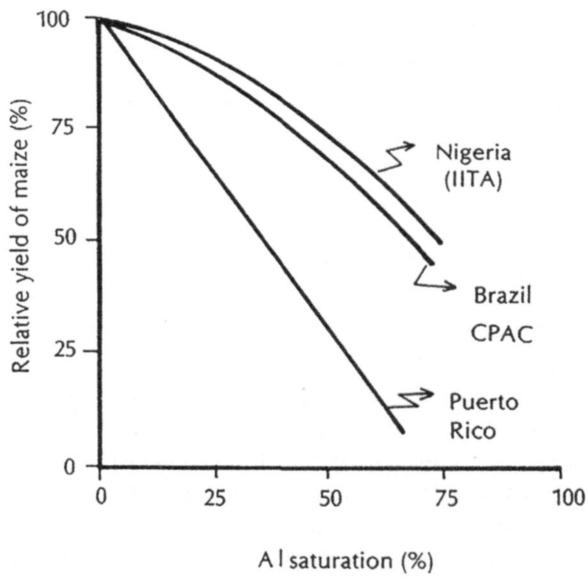

Fig. 3.17 Relationship between the relative yield of corn and aluminium saturation from three experiments on Oxisols 1 (Brazil), Oxisol 2 Puerto (Rico), and Alfisol (Nigeria). (Source: Collated by Bouldin et al. 1987)

Even in the absence of the latter sources, sodicity generally increases in subhumid and semi-arid Vertisols after these are cleared of natural vegetation for cropping. The increase in ESP is more rapid in cultivated Vertisols with low clay content than with high clay content (Table 3.10). However, the adverse effect of a given amount of exchangeable Na (at a similar salt concentration) on soil permeability, as measured by dispersed clay, increases with the increasing clay content (Fig. 3.17). Thus, even at low exchangeable Na (ESP, 1–10), soil physical conditions and hence soil water relationships in Vertisols can be adversely affected. Therefore, the ESP limit of sodicity set for the coarse-textured soils, >15 (USSL 1954), is not valid for the fine-textured Vertisols, and the sodicity effect is considered at ESP >6 for these soils (Loveday 1976).

The sodicity causes several adverse effects on soil properties as well as plant growth (Fig. 3.18).

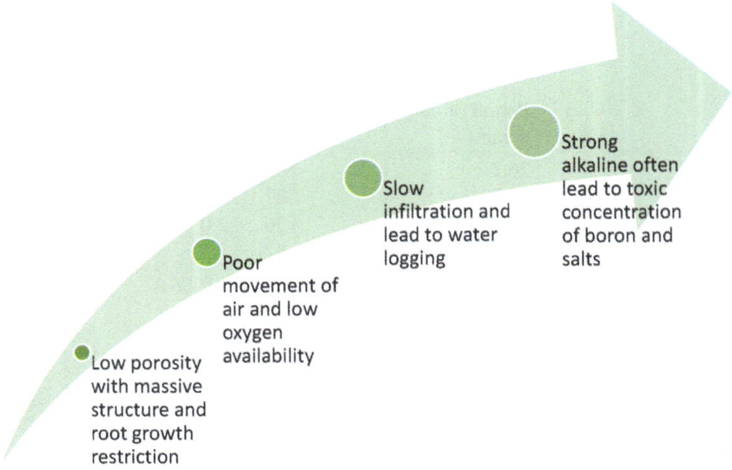

Fig. 3.18 Problems associated with soil sodicity. (Modified from Rengasamy 2002; Rengasamy et al. 2010)

The effect of sodicity on soil dispersion and other adverse physical properties such as crust formation (Shainberg 1985) is accentuated in smectite-dominant soils such as Vertisols, primarily because of the swelling characteristics of smectites with increasing exchangeable Na, thus resulting in reduced soil permeability (McNeal and Coleman 1966; McNeal et al. 1966; Cass and Sumner 1982; Coulombe et al. 1996, 2000). Dispersion and subsequent translocation of both smectites and other clays containing exchangeable Na block the water transmission pores even further (Rengasamy et al. 1984; Coulombe et al. 1996; Jayaraman et al. 2018). As a result, at a given ESP (and EC), the infiltration rate decreases the most in smectitic soils and the least in kaolinitic soils (Levy and van der Watt 1988). However, both swelling and dispersion characteristics of such systems can be moderated by maintaining the electrolyte concentration of the soil solution above a threshold value (Cass and Sumner 1982; Rengasamy et al. 1984; Quirk 1986). Thus, the amount of clay dispersed in the soil is not only affected by exchangeable Na and clay content (and clay mineralogy), as shown in Fig. 3.19, but also by the composition and concentration of the soil solution (SAR, EC). That is,

Dispersed clay (g/100 g) = 0.74 + 1.09 exch. Na (cmol kg^{-1}) + 0.06 clay

$$\text{content }(\%) - 9.8\ \text{EC}\ \left(\text{dSm}^{-1}\right) \qquad (3.23)$$

$$r^2 = 0.71, n = 119$$

or Dispersed clay (g/100 g) = 0.20 + 0.40 ESP + 0.04 clay content (%) + 0.05 CEC

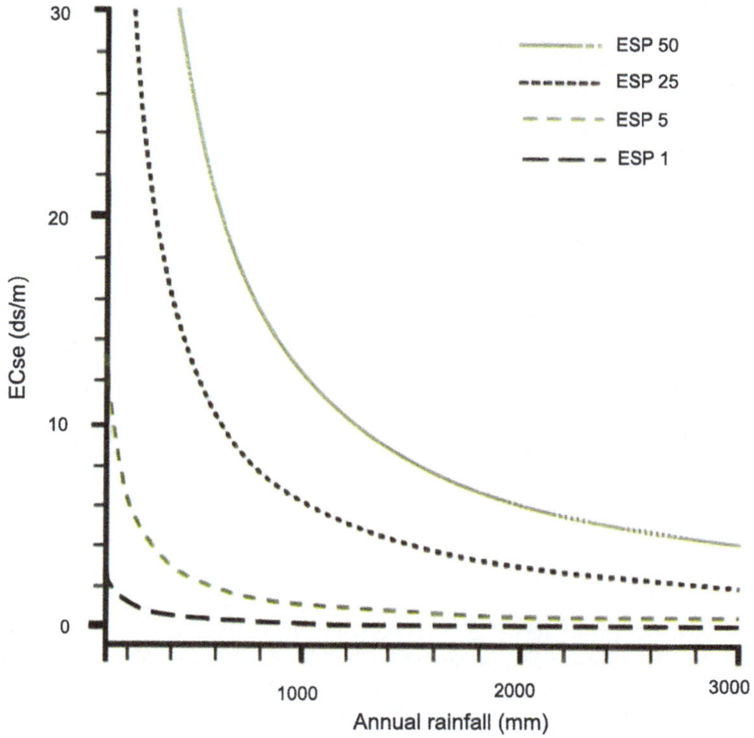

Fig. 3.19 Interrelationship among EC and ESP of soil (kaolinite-smectite and illite-smectite mineralogy) and mean annual rainfall, in tropical and subtropical regions. (Source: Salinity Management Handbook 2011)

$$\left(\text{cmol kg}^{-1}\right) - 12.5\ \text{EC}\ \left(\text{dSm}^{-1}\right) \tag{3.24}$$

$$r^2 = 0.72, n = 119$$

The concept of threshold electrolyte concentration (Quirk and Schofield 1955; Shainberg and Letey 1984) or total cation concentration (Rengasamy et al. 1984) to manage sodicity and hence soil permeability emphasises the management of hydrologic cycle to avoid excessive leaching of soluble salts on the one hand and reduce evaporation and hence accumulation of excessive soluble salts on the other, thus avoiding the adverse effect of sodicity (and salinity) on crop production (Maas 1986). Except in highly sodic soils containing low electrolyte concentration in soil solution where amendments such as the addition of gypsum may be required, biological management of sodicity through optimum crop production and subsequent soil surface management, including crop residue retention and minimum tillage (Dalal 1989), provide the long-term solution to arrest or even reverse soil sodicity as a process of soil fertility degradation.

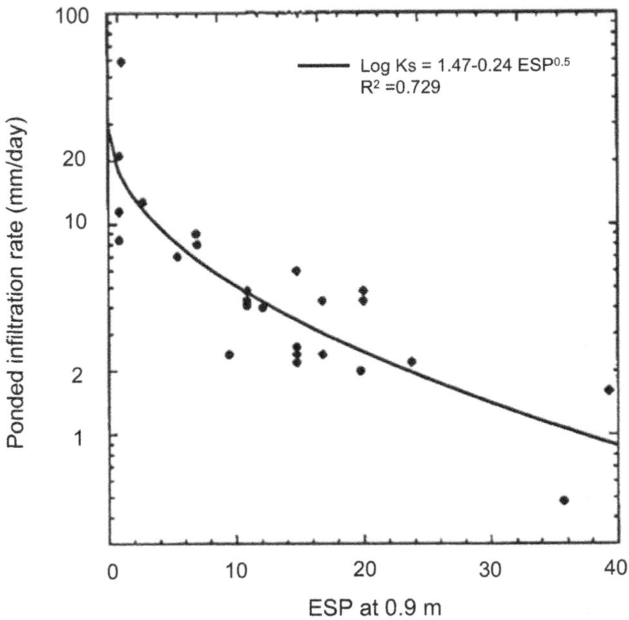

Fig. 3.20 The relationship between ponded infiltration rate (1.5 to 7 days) and exchangeable sodium percentage (ESP) at 0.9 m depth for soils of 25 to 82% clay content from the field data of Shaw and Yule (1978) and Gardner and Coughlan (1982). (Source: Shaw 1995)

3.5.2.2 Sub-soil Sodicity

High sub-soil sodicity (and salinity) is often more widespread than the surface sodicity in many Vertisols (Probert et al. 1987; Naidu and Sumner 1998. Rengasamy et al. 2022) and calcareous Alfisols and Inceptisols in the subhumid and semi-arid tropical and subtropical regions. In fact, the infiltration rate in the field and saturated hydraulic conductivity of the whole profile may be limited by the sub-soil sodicity (Fig. 3.20) which is also generally accompanied by high sub-soil salinity because of the extremely slow leaching of salts from these layers in the semi-arid environments.

For example, in Vertisols by increasing the ESP from 5 to 10, plant available water capacity (PAWC) is decreased from 300 to 220 mm, and by further doubling the ESP from 10 to 21, the PAWC decreased from 220 mm to only 140 mm (Table 3.11). Similar large reductions in PAWC from an increase in ESP were also observed in Alfisols in the semi-arid regions (Gardner and Coughlan 1982).

Sub-soil sodicity presents an actual and potential hazard to crop productivity; it could be the main course of poor hydraulic conductivity in soil and consequently poor plant available water capacity, excessive runoff and soil erosion, salt scalds, waterlogging, and salt pans. It is also a potential source of sodification (and salinity) of soil under irrigation (Dowling et al. 1984) and a source of sodicity in the recharge areas.

Table 3.11 Plant available water capacity (PAWC) and exchangeable sodium percentage (ESP) at 0.5–0.6 m depth for five soils in sub-tropical Queensland (mean annual rainfall, 700 mm)

Soils	ESP (0.5–0.6 m)	PAWC (mm)
Alfisol 1	1	190
Alfisol 2	36	88
Vertisol 1	5	300
Vertisol 2	10	220
Vertisol 3	21	140

Source: Gardner and Coughlan (1982), Agric Chem Branch Report No. 20, Department of Primary Industries, Brisbane, the State of Queensland

Sodic soils are distributed widely in both cultivated and grazing lands, in dryland and irrigated areas, and from subhumid to semi-arid and arid regions with increasing salinity hazards as well (Fig. 3.20). Unfortunately, the exact extent of sodic-affected areas of cultivated lands is not well delineated in tropical and subtropical regions. Szabolcs (1979) estimated that the total areas affected by sodicity were 59.5 Mha in South America, 26.9 m Mha in Africa, 1.8 Mha in South Asia, and 340 Mha in Australia. Later, Montanarella et al. (2015) estimated that the total areas affected by sodicity were 129.3 Mha in South America, 86.7 Mha in Africa, 1.8 Mha in South Asia, 120.2 Mha in North and Central Asia, 9.6 Mha in North America, and 340 Mha in Australia. Of the total affected area in Australia, about 0.82 Mha of cultivated land is classified as salt-affected (Peck 1981). Only a small proportion of that area, mostly Alfisols and Vertisols, is sodic in the top layer, although a large proportion of the area is affected by sub-soil sodicity and alkalinity (Shaw et al. 1987). Recent estimates suggested that a total area of 835 Mha is affected by saline and sodic problems, in which 399 Mha and 436 Mha are affected by saline and sodic soils, respectively (FAO Land and Plant Nutrition Service 2008; Leogrande and Vitti 2019).

3.5.2.3 Consequences of Sodicity

In Alfisols and Inceptisols and especially in Vertisols, adverse effects of sodicity have been observed with a surface layer ESP of 1 to 1.5 on crusting (Shainberg 1985) and an ESP of >3 on dispersion (Rengasamy et al. 1984) and infiltration (Shaw and Thorburn 1985), although the magnitude of adverse effects of ESP is modified by total salt concentration in the soil solution, clay content, clay mineralogy, and organic matter.

High levels of sodium in the soil affect crop growth directly by accumulating in the leaves causing leaf burn and even defoliation and indirectly through deteriorated soil-water and air permeability, thus resulting in poor root development. In sodic soils with ESP >50 and CO_2 partial pressure of <0.01 atmosphere, Ca concentration in soil solution becomes limiting (<0.4 moles L^{-1}) for plant growth (Wallace 1966), and concomitantly, bicarbonate concentration interferes with plant growth and nutrient uptake and especially Fe uptake. On the other hand, toxic levels of chloride, boron, aluminate, and often molybdenum may also limit crop growth. Pearson (1960) and Gupta and Abrol (1990) have collated the relative crop tolerance to soil sodicity.

Sodicity can be managed by appropriate management practices involving sodic tolerant crop varieties and the application of gypsum along with organic manures to

Table 3.12 Dispersive behaviour or sodicity rating of soil

Details	Slight	Moderate	Severe
Exchangeable sodium percentage (ESP %)	6–10	10–15	>15
Gypsum application rate on neutral to acid soils ($Mg^{-}ha^{-1}$)	0–1.5	2.5	5
Gypsum application rate on alkaline soils ($Mg\ ha^{-1}$)	1.0–2.5	5	5.03

Source: Modified from Rengasamy et al. 2010

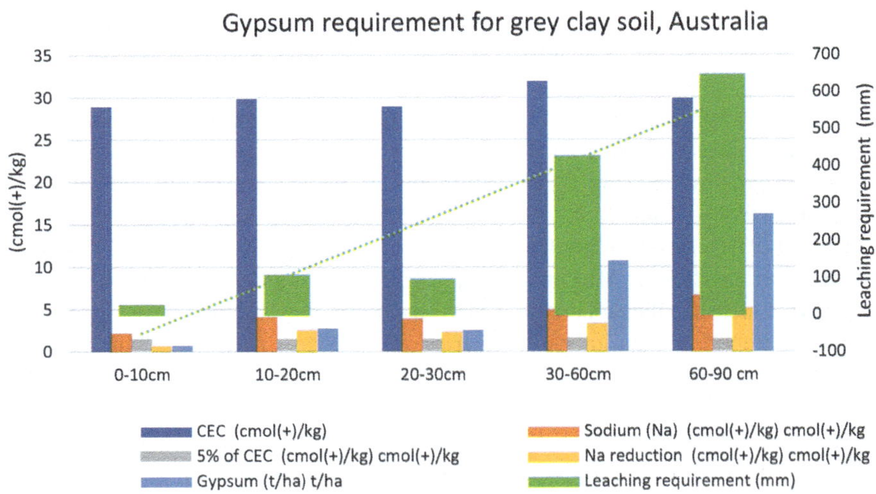

Fig. 3.21 Gypsum and leaching requirement of grey clay soil (Vertisol), Australia. (Source: Modified from Menzies et al. 2019)

overcome these problems. Rengasamy et al. (2010) have reported that gypsum requirement can be quantified by assessing the dispersive nature of soil (Table 3.12). Similarly, Menzies and co-workers (2019) have pointed out that at a mechanistic level, two processes, namely, 'swelling and dispersion' are responsible for the behaviour of sodic soils. Moreover, these two processes are controlled by the soil surface charge as well as exchangeable cations. Menzies et al. (2019) reported that gypsum application can be very effective in improving soil surface conditions, providing better soil tilth, and reducing surface crusting. They also quantified the gypsum requirement as well as the leaching requirement of grey clay soil (Vertisol) in Australia (Fig. 3.21).

Due to the increased cost of amendments (particularly gypsum, gypsum slurry, deep ripping, and placement of gypsum in sub-soil), the application of organic amendments with gypsum or in saline-sodic soil elemental S provides another option (Dang et al. 2010; McBeath et al. 2010; Armstrong et al. 2015; Leogrande and Vitti 2019). Dang et al. (2010) reported that a one-time surface applied gypsum at 2.5 Mg ha^{-1} increased cumulative gross margins by \$207 ha^{-1} over 4 crops (wheat, chickpea, wheat, and sorghum), removed 115 Mg ha^{-1} sodium chloride in the rooting zone, and improved PAWC by 15 mm. Further, organic amendments with gypsum remediate sodicity as well as improve soil aggregation, thereby increasing both the infiltration rate and the hydraulic conductivity of soil (Niaz et al. 2022).

Elemental S application to an alkaline-sodic soil is also used to remediate sodicity, since added S when oxidised by microbes in soil produces H_2SO_4, which dissolves $CaCO_3$ granules and releases Ca^{2+} for replacement of Na^+ from the exchange sites (Osman 2018). No-till and conservation agriculture practices including no-till, crop rotations including legume, and crop residue retention lead to a decrease in alkaline soil pH, as well as a decrease in exchangeable ESP (Dalal 1989; Jat et al. 2024) across alkaline-sodic soils including Vertisols and Inceptisols/Entisols in the subtropical and tropical regions.

3.5.3 Excess Element Load

3.5.3.1 Salinity

Soil fertility degradation is caused by an excess load of nutrients and other elements, collectively measured by electrolyte concentration and electrical conductivity (EC), that is, salinity (Rengasamy et al. 2010; Leogrande and Vitti 2019). FAO (1979) classified soil fertility degradation due to salinity into four groups on the basis of increase in EC in the 0–0.6 m layer: <2, 2–3, 3–5, and 7.5 dS m^{-1} $year^{-1}$, respectively, and the corresponding critical status are none to slight, moderate, high, and very high risk to salinity. As shown in Fig. 3.22 (Sect. 3.3), not only does the risk to salinity increase as the rainfall decreases, especially under inadequate supplementary irrigation (< leaching fraction) or when salty groundwater is present near the surface, but also the sodicity risk further increases as the clay content increases.

NRCS Soil Survey handbook classified soil salinity based on the electrical conductivity (1 dS m^{-1} = mmhos cm^{-1}), namely, non-saline (0 < 2), very slightly saline (2 < 4), slightly saline (4 < 8), moderately saline (8 < 16), and strongly saline

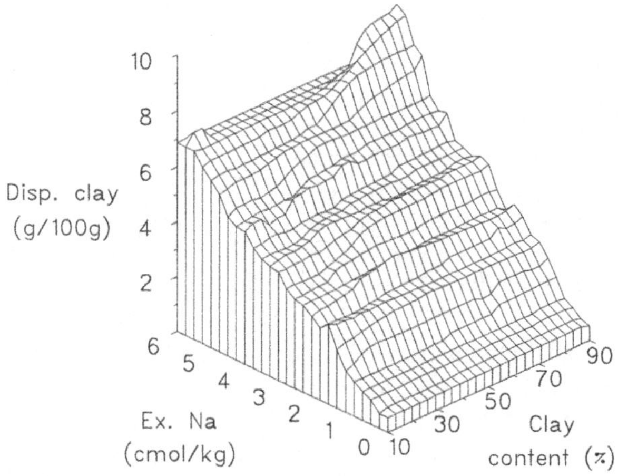

Fig. 3.22 Effect of exchangeable Na on clay dispersion depending on the clay content of the soil. The figure is drawn from the relationship: Dispersed clay (%) = −0.23 + 0.64 exch. Na + 0.065 clay content (%), $r^2 = 0.63$, $n = 119$ (Alfisols and Vertisols). (Source: R.C. Dalal, unpublished data)

Table 3.13 The relative salt tolerance of a range of plants expressed in order of salinity threshold (100% relative yield)[a]

Plant	Salinity threshold (dSm^{-1})	Rate of yield decrease (dSm^{-1} % $yield^{-1}$)	References[b]
Carrot (Daucus carota)	1.0	14.1	1
Silverload desmodium (Desmodium	1.0	15.2	2
uncinatum)	1.0	18.9	1
Mungbean (Phaseolus vulgaris)	1.0	33.3	1
Strawberry (Fragaria)	1.1	6.9	5
Brinjal (Solanum melongena)	1.2	16.1	1
Onion (Allium cepa)	1.3	2.1	4
Avocado (Persea americana)	1.3	14.3	1
Cowpea (Vigna ungiuculata)	1.5	10.3	2
Berseem (Trifolium alexandrinium)	1.5	11.1	1
Sweet potato (Ipomoea batatas)	1.7	5.9	1
Sugarcane (Saccharum officinarum)	1.7	12.0	1
Corn (Zea mays)	1.7	12.0	1
Flax (Vinum usitatissimum)	1.7	12.0	1
Potato (Solanum tuberosum)	1.8	9.0	2
Paspalum (Paspalum dilatatum)	2.0	4.0	2
Pangola grass (Digitaria decumbens)	2.0	7.3	1
Lucerne (Medicago sativa)	2.0	7.9	2
Siratro (Macroptilium atropurpureum)	2.3	7.0	1
Sesbania (Sesbania exaltata)	2.4	20.4	2
Townsville stylo (Stylosanthes humilis)	2.8	4.3	1
Sudan grass (Sorghum sudanense)	3.0	3.0	2
Kikuya grass (Pennisetum clandestinum)	3.0	12.2	1
Rice (Oryza sativa)	4.0	3.4	1
Date (Phoenix dactylifera)	5.0	20.0	6
Oats (Avena sativa)	5.0	20.0	1
Soybean (Glycine max)	5.5	20.0	6
Sunflower (Helianthus annus)	6.0	6.8	2
Buffel grass (Cenchrus ciliaris)	6.0	7.1	1
Wheat (Triticum aestivum)	6.8	15.9	3
Sorghum (Sorghum bicolor)	6.9	6.4	1
Couch grass (Cynodon dactylon)	7.0	3.2	2
Rhodes grass (Chloris gayana)	7.7	5.2	1
Cotton (Gossypium hirsutum)	8.0	5.0	1
Barley (Hordeum vulgare)	8.5	12.4	2
Urochloa (Urochloa mosambicensis)	15.0	3.8	1
Guayule (Simmondsia chinensis)			

[a]Relative yield (%) = 100 – B (ECse – A), where A is the salinity threshold (dSm^{-1}), and B is the rate of relative yield (%) decrease (dSm^{-1} % rel. $yield^{-1}$)
[b]Sources: 1. Maas and Hoffman (1977). 2. Russell (1976). 3. Maas (1986). 4. Ayers (1977). 5. Heuer (1986). 6. Shaw et al. (1987)

(>16). Though the soil EC does not directly affect crop growth, it has been used as a criterion/indicator to assess the amount of nutrients available to crops (USDA_NRCS 2011). The decline in crop productivity due to salinity varies widely depending on the crop (Table 3.13), crop cultivar (genotypic variability), stage of crop growth, soil moisture, environmental factors (temperature, humidity, and light intensity), and other soil fertility constraints. For different crops, Smith and Doran (1996) reported that yield reduction occurred from 5.0% (barley) to 29% (peanut) under varying levels of threshold EC value at 1:1 (dS m^{-1}).

Similar to sodicity, the extent of soils degraded by salinity is not accurately known. Szabolcs (1979) estimated 2 Mha in Mexico and Central America, 69 Mha in South America, 53 Mha in Africa, 83 Mha in South Asia, 20 Mha in Southeast Asia, and 17 Mha in Australia. In semi-arid tropical and subtropical regions, Vertisols account for a large proportion of the sodicity-affected soils. Recent estimates reported that Asia, the Pacific, and Australia region account for the largest sodicity problem in the world (249 Mha, 8%) followed by Europe (73 Mha, 3.6%) and Latin America (51 Mha, 2.5%) (FAO Land and Plant Nutrition Service 2008; Leogrande and Vitti 2019).

3.5.3.2 Accumulation of Essential Nutrient Elements in Toxic Amounts

The mention was made of boron accumulation reaching toxic amounts in the processes of salinisation and alkalisation in the semi-arid tropics. So far, it appears that accumulation of B in toxic amounts occurs in sub-soil layers of Alfisols (Cartwright et al. 1984, 1987; Moraghan et al. 1991), and only limited areas of Vertisols are known to contain toxic levels of B. Suriyagoda et al. (2022) found that soils registered high concentration of essential mineral (N, P, K, Cu, Zn, and Mn) and toxic trace elements (As, Cd, and Pb) in a high rainfall and mountainous regions of Sri Lanka. Some of the micronutrients are very much essential for plant growth; however, some of the micronutrients are considered toxic under contrasting soil pH (boron accumulation to toxic levels under high pH in dry environments of alkaline soils) (Langridge 2022).

Since Cu, Zn, Fe, Mn, and Mo are required by plants in small amounts, high application of these elements, either through fertilisers, smelters, waste products, or industrial and municipal sewage sludge, leads to soil fertility degradation. Logan and Chaney (1983) reported the range of concentrations of Cu (84–17,000 mg kg^{-1}), Zn (101–49,000 mg kg^{-1}), Fe (1000–154,000 mg kg^{-1}), Mn (32–9870 mg kg^{-1}), and Mo (0.1–214 mg kg^{-1}) in the dry digested sewage sludges. High application rates of sewage sludges containing these nutrient elements at high concentrations are likely to be phytotoxic (Pendias and Pendias 1992; Pendias 2000, 2004; Jayaraman et al. 2005, 2011, 2016). For example, McGrath et al. (1988) stated that the maximum permissible concentrations of Zn in cultivated soils treated with sewage sludge were 200, 250, 300, and 450 mg Zn kg^{-1} dried sludge at pH 5.0–5.5, 5.5–6.0, 6.0–7.0, and >7.0, respectively; the corresponding values for Cu were 80, 100, 135, and 200 mg Cu kg^{-1} dried sludge.

Soler-Rovira et al. (1995) studied the critical limit values of heavy metals in soils and sewage biosolids with respect to pH and proposed the critical concentration of heavy metals. They also defined the limit values of heavy metal concentration in soil. Treated sewage biosolid may not be applied to soils having higher limits of concentration so that the values do not exceed the limit values as proposed above. However, it should not exceed the limit values set in accordance with the heavy metal content of the soil and sewage biosolids to be used for soil application (Jayaraman

2001; Jayaraman et al., 2005, 2011, 2016). Kabata-Pendias and Pendias (1992) identified typical concentrations of trace metals, namely, deficient, sufficient or normal, excessive, or toxic and tolerable to agricultural crops in plant leaves.

3.5.3.3 Accumulation of Non-essential Elements in Toxic Amounts

Toxic concentrations of Al and Mn accumulate or are produced in Ultisols, acidic Alfisols, and Inceptisols. This has been discussed in Sect. 3.5.1.

Many Alfisols and Ultisols require heavy applications of phosphatic fertilisers for maintaining optimum concentrations of P for crop growth. Since most P fertilisers also contain Cd, 400 ± 40 mg Cd kg^{-1} P, Rayment et al. (1988, 2005) reported that a single heavy application of P fertilisers may result in excessive accumulation of Cd in soil, about 2.5–10 mg kg^{-1} soil above the background soil values (Logan and Chaney 1983). In addition, the excessive application of phosphatic fertiliser results in the accumulation of arsenic (As) and lead (Pb) in soil eventually transferred to plants and animals (Jiao et al. 2012). Besides, rock phosphates also contain trans-uranium elements, such as U and Th, which even at small concentrations are not only phytotoxic but also radiation toxic to animals and humans. Fortunately, these elements are strongly sorbed by soil constituents, although regular monitoring of soil and crops is advised where heavy rates of phosphate rock applications are advocated, such as in Ultisols and in some Alfisols and Oxisols (Sanchez 1987; Nziguheba and Smolders 2008).

Land disposal of mine spoils, industrial wastes, and sewage sludge carries the risk of the accumulation of As, Cd, and heavy metals such as Pb, Hg, Cr, Co, and Ni to phytotoxic levels, especially in acidic Alfisols and Ultisols, and consequently a hazard to animal and human health (He et al. 2005; Jayaraman et al. 2005). The reported ranges of concentration of As, Cd, Pb, Hg, Cr, and Ni in the digested sewage sludge are, respectively, 1.1–230, 1–3410, 13–26,000, 0.6–56, 10–99,000, and 2–5300 mg kg^{-1} dry sludge (Logan and Chaney 1983). Once ingested, all of these metals accumulate/bio-transfer into soil-plant-animal/human system (Jayaraman et al. 2005), and therefore, continual disposal of wastes containing heavy metals can lead to long-term soil fertility degradation which may be extremely slow to reclaim (Logan 1992) and prohibitively expensive.

On the other hand, most of the xenobiotic chemicals (herbicides and pesticides) applied to crops and soils are chemically and biologically degraded into non-active and neutral products (Hicks et al. 1990; Wang et al. 2022). Most herbicides and pesticides do not appear to have any long-term harmful effects on soil microbial activity when these are applied at recommended field levels (Smith 1982; Smith and Novak 1987), although at higher rates, microbial activity may temporarily be inhibited (Young 1984; Wang et al. 2022). However, from a human health perspective, minimal and prudent use of xenobiotics in soil–plant system is imperative (Katayama et al. 2010; Wang et al. 2022).

3.6 Aggregate Breakdown

3.6.1 Aggregate Formation and Stabilisation in Soil

The arrangement of individual soil particles (i.e. sand, silt, and clay) into larger units is known as '*soil structure*' and generally called '*aggregates*'. Units of soil structure are called '*peds*' which occur through natural processes, whereas '*clods*' occur through disturbances such as tillage practices and other agricultural management. Soil structure has been classified into three distinct groups, namely, (i) structureless (e.g. single grain), (ii) massive structure (e.g. granular, platy, blocky and prismatic), and (iii) structure destroyed (e.g. puddled soils—ploughed under very wet condition) (Troeh and Thompson 2005a, b). According to Martin et al. (1955), a soil aggregate is a naturally occurring cluster or group of soil particles in which the forces holding the particles together are much stronger than the forces between adjacent aggregates (Hillel 2007). Physical forces predominate in the formation of soil aggregates. These are primarily wetting and drying, freezing, and thawing (temperate regions and some tropical and subtropical highlands), the compressive and drying action of roots, and the mechanical and biochemical alimentary action of soil fauna (mainly earthworms, termites, and ants). Under the influence of these forces, the primary particles are brought close together so that in the presence of aggregate-stabilising agents, the van der Waals forces, hydrogen bonding, short-range electrostatic bonding, ligand exchange, covalent bonding, and a combination of two or more of these linkages in organics-polyvalent, metal cations-clays, lead to the formation and stabilisation of soil aggregates (Dalal and Bridge 1996; Hillel 2007). The hydrophobicity of some aggregate-stabilising agents also confers stability against water entry into aggregates. The soil aggregate-stabilising agents include SOM and organic materials of plant, microbial and animal debris; root exudates (organic acids/enzymes) and products of microbial synthesis; plant roots (Oades and Waters 1991), fungal hyphae, and microorganisms and their products (Tisdall 1994); and various inorganic substances, clays, polyvalent metal cations, Ca^{2+}, Fe^{3+} and Al^{3+}, oxides of iron, aluminium, manganese and silica, and calcium carbonate (Brady 1996; Hillel 2007).

Tisdall and Oades (1982) suggested a hierarchical organisation of soil aggregates, from <2 μm, 2–20 μm, 20–250 μm, 250–1000 μm, to >1000 μm, mainly based upon their results on water stability of soil aggregates for Alfisols. They presented a model of the organisation of soil aggregates along with their stabilising agents, both organic and inorganic (Fig. 3.23). The organic agents for stabilising the aggregates were classified into three groups: (1) transient—mainly polysaccharides; (2) temporary—mainly roots and fungal hyphae; and (3) persistent—resistant aromatic components associated with polyvalent metal cations and strongly sorbed polymers. The hierarchal organisation of soil aggregates is probably not valid for all soils. For example, in Vertisols, macroaggregates >1000 μm diameter generally break down to < 30 μm diameter aggregates (Collis-George and Lal 1970) and in Oxisols to < 20 μm diameter aggregates, silt and clay particles (Oades and Waters 1991), and there was no trend of aggregate hierarchy. On the other hand, Alfisol and

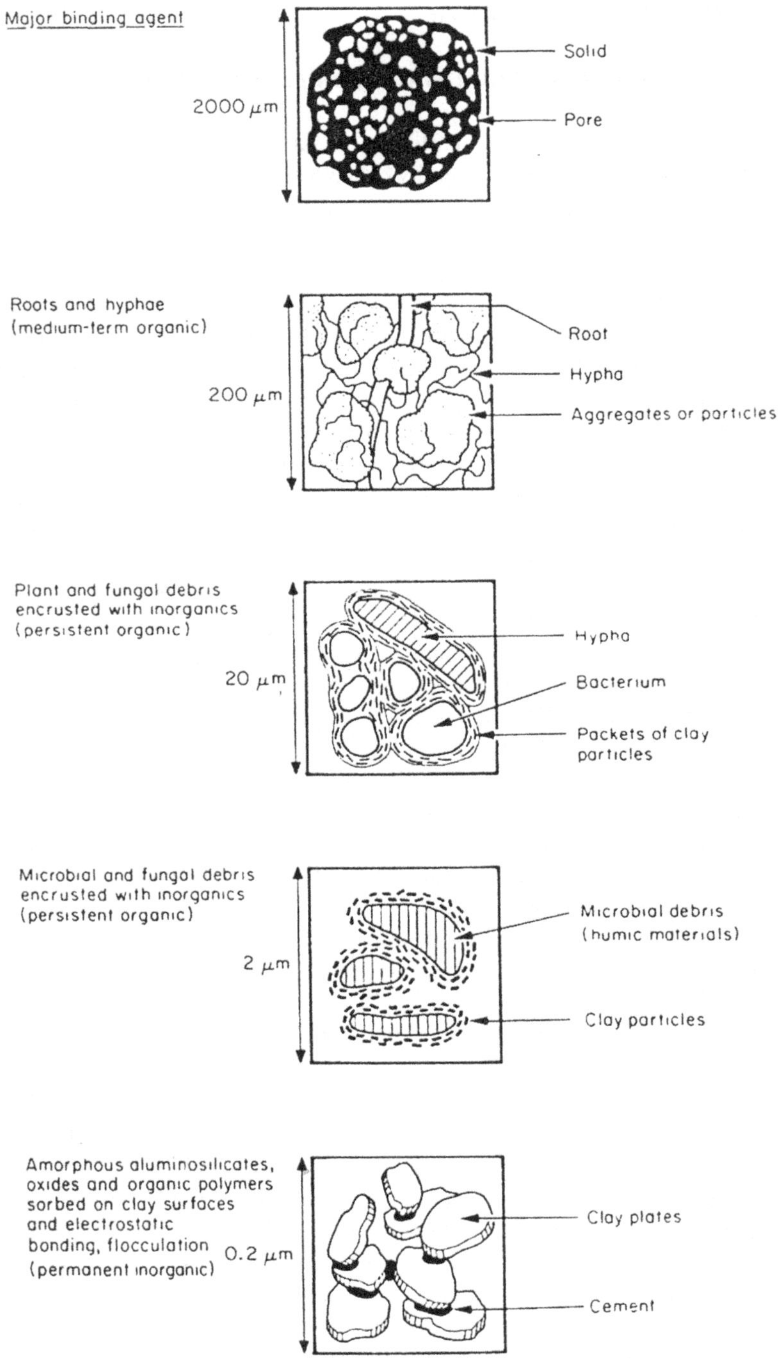

Fig. 3.23 Model of a soil aggregate showing the operation of different organic components at various levels of aggregate size (adapted from Tisdall and Oades 1982). (Source: Theng 1987)

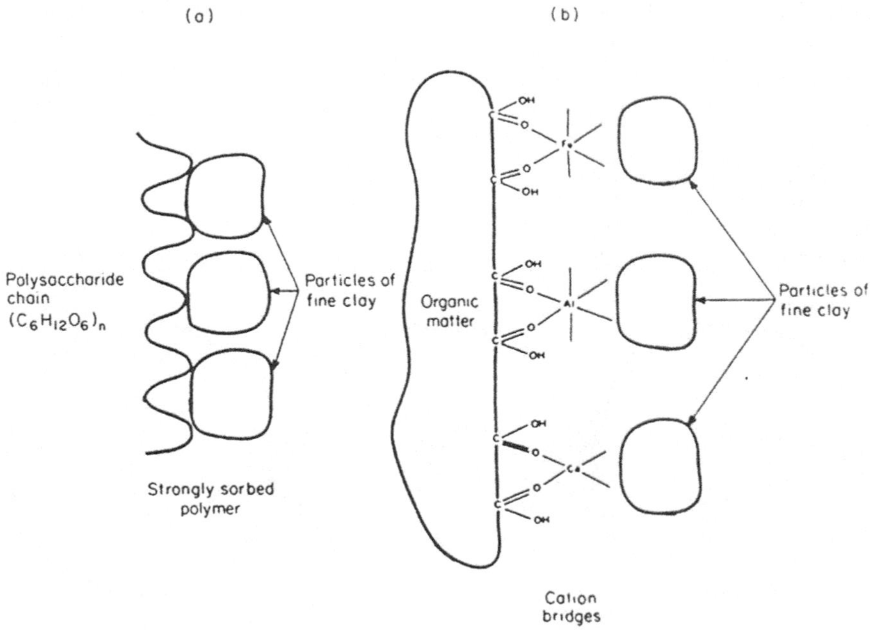

Fig. 3.24 Diagram showing some of the possible bonding modes between organic, exchangeable divalent m^{2+} cations and clay surfaces: (**a**) cation bridging through water molecules giving rise to an outer-sphere complex; (**b**) direct cation bridging giving rise to an inner-sphere complex; (**c**) anion-exchange or electrostatic interaction; (**d**) ligand-exchange or specific adsorption. The possibility of H bonding is also indicated in schemes (**a**) and (**d**). (Sources: Theng 1987)

Mollisol tend to break down in steps, namely, macroaggregates >250 μm diameter break down into microaggregates (20–250 μm diameter) before particles break down into <20 μm (Oades and Waters 1991).

Macroaggregates, >250 μm diameter (Edwards and Bremner 1967), are stabilised by roots and hyphae (Tisdall and Oades 1982; Oades and Waters 1991) and macroorganic matter (Coughlan et al. 1973; Elliott 1986). These aggregate-stabilising agents and hence macroaggregates are largely affected by crop (for example, crop species and crop sequences especially roots and rhizosphere exudates and microbial products) and soil management (for example, crop cover, residue cover, tillage operations, and machinery) (Dalal and Bridge 1996, Jayaraman et al. 2017, Jayaraman et al. 2018a, b, c, d; Sarkar 2018; Hati et al. 2021). On the other hand, the water stability of macroaggregates, <250 μm (Edwards and Bremner 1967), is largely due to persistent organic binding agents (Tisdall and Oades 1982), polyvalent metal cations, oxides of iron and aluminium, calcium carbonate, and salt bridges (Fig. 3.24), and hence, it is less affected by crop and soil management.

Table 3.14 Aggregate class and size, approximate equivalent pore diameter and soil water matric potential

Aggregate		Pore diameter (μm)	Pore water suction (kpa)
Class	Diameter (μm)		
Clay-size	<2	<0.3	>980
Microaggregates	2–20	0.3–3	98–980
Microaggregates	20–250	3–39	7–98
Microaggregates	250–1000	39–155	1.9–7
Microaggregates	>1000	>155	<1.9

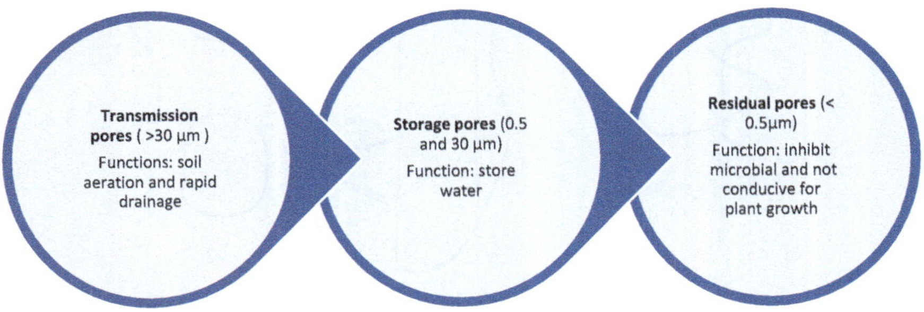

Fig. 3.25 Pore size and its functions in soil (Source: Authors)

3.6.2 Functions of Soil Aggregates

The size, arrangement, and stability of soil aggregates and the pores contained within intra- and inter-aggregates affect water retention and movement (Table 3.14), soil air, and hence crop growth. For example, pores >39 μm (macroaggregates, >250 μm) transmit water for recharge of water-depleted soil profile and drainage, and pores between 0.3 and 29 μm (microaggregates, 2–250 μm) retain water for crop growth (Table 3.14) (Greenland 1977; De Leenheer 1977; Oades 1984; Dalal and Bridge 1996). In a well-aggregated fertile soil, a combination of aggregate sizes and hence soil pores assure adequate drainage and plant available water retention for crop growth, good aeration, and root growth (Rowell 1994; Kay and Vandenbygaart 2002; Nimmo 2003; Lal and Shukla 2004). Soil pores are broadly grouped into three categories according to the pore size and pore functional characteristics of soil, namely, transmission pores, storage pores, and residual-bonding pores (Lal et al. 2004; Nimmo 2003; Hillel 2003). In addition, based on the pore diameter, it has been classified into macropore, mesopore, and micropore, and often, this classification has been used as a soil quality indicator (Pagliai and Vignozzi 2002, Nimmo 2003; Azizsoltani et al. 2019). Moreover, aggregates are denser than the structured soil (bulk), and the intra-aggregate pore size distribution not only has more fine pores but also a more tortuous path (Horn 1990; Bhattacharya et al. 2021) (Fig. 3.25).

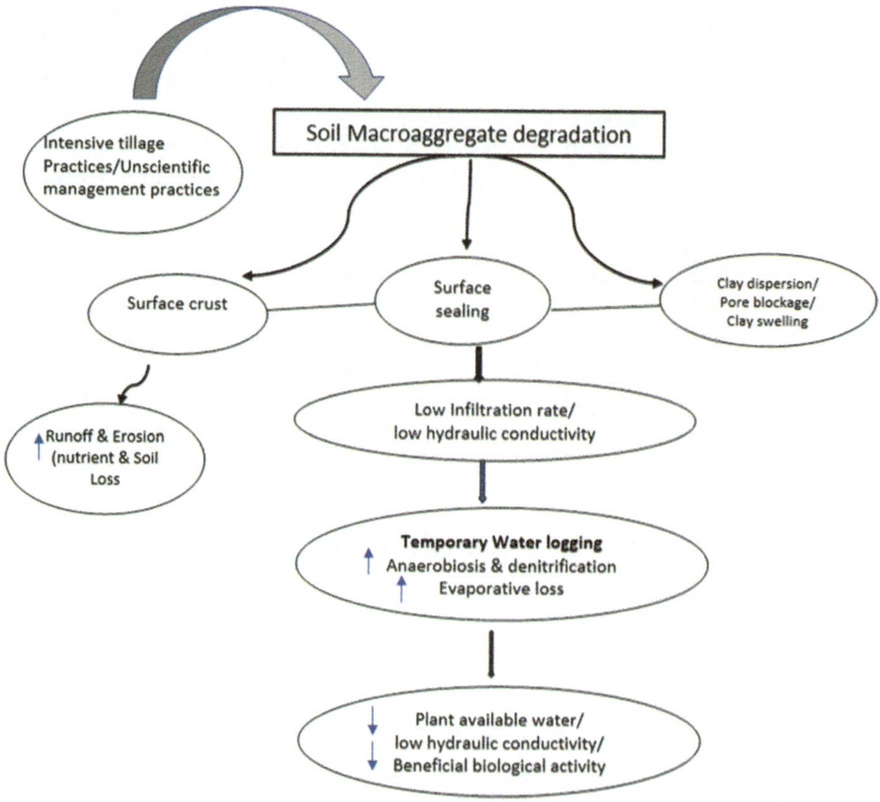

Fig. 3.26 Effects of soil aggregate breakdown on soil fertility degradation through a number of interrelated degradative processes (runoff and erosion, nutrient loss, poor air-water relationships, and reduced plant available water stored in the soil profile). Source: extended and adapted from Coughlan (1984) and Cook (1988) and modified from Dalal and Jayaraman (2021)

3.6.3 Consequences of Soil Aggregate Breakdown

The effects of soil aggregate breakdown on soil fertility degradation are exhibited through a number of interrelated degradative processes, arising mainly from surface instability (sealing and crusting, dispersion, blocked pores >39 μm, low infiltration rate, and hydraulic conductivity) and surface and sub-surface compaction (temporary waterlogging). Some of the soil fertility degradative processes accelerated by aggregate breakdown are outlined in Fig. 3.26. These are soil erosion and runoff, accompanied by nutrient and soil loss (Sect. 3.1); anaerobiosis, accompanied by denitrification (Sect. 3.4) and temporary accumulation of toxic substances; SOM loss (Craswell et al. 1972; Dalal and Mayer 1986a; see Sect. 4.1), salinity and sodicity, and sub-soil salt accumulation in semi-arid tropical and subtropical regions (Sect. 3.5); and hard-setting of soils (Mullins et al. 1990) containing non-swelling clays (for example, many Alfisols and Inceptisols). As a consequence, plant available water in the soil profile is greatly reduced, and available nutrients and beneficial biological activity are also reduced so that crop production is likely to be affected. Unfortunately, it has been difficult to establish relationships between the extent of aggregate breakdown (for example, water-stable aggregates) and crop yields (Low 1972), primarily because of a number of soil fertility degradative processes affecting crop growth simultaneously, although crop root growth is substantially reduced (El-Swaify et al. 1985).

It has been frequently shown, however, that the steady infiltration rate and saturated hydraulic conductivity decline with the decrease in mean weight diameter and proportion of water stable aggregates >20 μm in Vertisols (Krishnamoorthy et al. 1966; Smith et al. 1984; Loch 1989; Cook et al. 1992), Alfisols (Lal 1984; El-Swaify et al. 1985), Oxisols (Silva et al. 2014), Ultisols (Seubert et al. 1977), and Inceptisols (Loch and Foley 1994). For example, Loch (1989) and Loch and Foley (1994) obtained the following relationships between steady infiltration rate and water-stable aggregates from Vertisols, Alfisols, Oxisols, and Inceptisols subjected to high energy rainfall:

$$Y = 57.4 \exp(-0.039\, x_1) \tag{3.25a}$$

$$Y = 76.4 \exp(-0.032\, x_2) \tag{3.25b}$$

$$Y = 52.2 - 0.6\, x_3 \tag{3.25c}$$

and

$$Y = 1.9 + 19.9\, x_4 \tag{3.25d}$$

where Y = steady infiltration rate (mm h^{-1}) of high energy rain (100 mm h^{-1})
x_1, x_2, and x_3 = percentage of soil aggregates <125 μm, <250 μm, and <500 μm, respectively
x_4 = mean weight diameter (mm)

Cook et al. (1992) found that as the mean weight diameter of a Vertisol under native grass decreased from 0.54 to 0.26 mm after 64 years of mechanical cultivation, the corresponding saturated hydraulic conductivity decreased from 14.4 mm h^{-1} to only 2.5 mm h^{-1}.

Aggregate stability has been assessed by Ribeiro et al. (2017) using ultrasonic energy. They found that Oxisols showed more aggregate stability compared to Ultisols of the Brazil region (Ribeiro et al. 2017). Aggregate breakdown not only affects soil physical properties but also significantly affects soil chemical and biological properties (Dalal and Bridge 1996, Jayaraman et al. 2017, Jayaraman et al. 2018a, b, c, d, 2024; Sarkar 2018; Hati et al. 2021).

Table 3.15 Water-stable aggregates in soils under natural vegetation or pastures and the adjacent cultivated soils

Soil	Period of cultivation (year)	Water-stable aggregates (%)		References
		Virgin	Cultivated	
Alfisol	16	62.7	45.4	Loch and Folley (1994)
Vertisol	6	87.0	81.0	Chan et al. (1988)
	8	85.0	80.0	Chan et al. (1988)
	22	84.0	76.0	Chan et al. (1988)
	45	85.0	77.0	Chan et al. (1988)
	55	80.8	78.0	Loch and Folley (1994)

3.6.4 Magnitude of Aggregate Breakdown in Cultivated Soils

Primarily, the aggregate breakdown is governed by soil texture, clay minerology, and the organic matter content of the soil; thus, it varies with soil types (Amézketa 1999; Six et al. 2000a, b; Nciizah and Wakindiki 2014; Xiao et al. 2017).

Mechanical cultivation for cropping of well-aggregated soils, such as those under pastures, virgin grasslands, and forests, leads to aggregate breakdown, especially in fallow-mono culture cropping systems (Table 3.15). Macroaggregates >250 μm in a Typic Pellustert declined from 55% under natural vegetation to 30% after 64 years of mechanical cultivation for cereal cropping (Fig. 3.27a, b). However, a much more rapid and larger decline in the mean weight diameter of soil aggregates occurred during this period (Fig. 3.27a, b). For example, the decline rate of macroaggregates >250 μm was 8.7% year^{-1} compared to that of a decrease in mean weight diameter, which was about five times faster (45.3% year^{-1}). The decline in macroaggregates >500 μm in virgin soil brought under cultivation is likely to be even much faster. Chan et al. (1988) found that a Vertisol (Ustic Chromuserts), which under native pasture had 60% macroaggregates >500 μm, contained only 3% of this size aggregates after 45 years of mechanical cultivation for cereal cropping.

Zanini et al. (1998) reported that when the aggregate stability has been measured under water-saturated conditions through conventional methods of wet sieving at a fixed time interval (i.e. 5 to 20 min), the dynamic features of aggregate breakdown are overlooked. Subsequently, they proposed an exponential equation to describe the aggregate breakdown process as a function of wet sieving time.

$$y(t) = a[1 - \exp(-t/c)] + b,$$

where

a = maximum estimated abrasion loss of aggregates
b = incipient failure of the aggregates when saturated in water
c = links the rate of aggregate breakdown to wet sieving time

The stability of water-stable aggregates under mechanical cultivation also varies with soil type, land use, and organic matter content and also due to variation in the nature and amount of aggregate-stabilising agents (Six et al. 2004; Jayaraman et al.

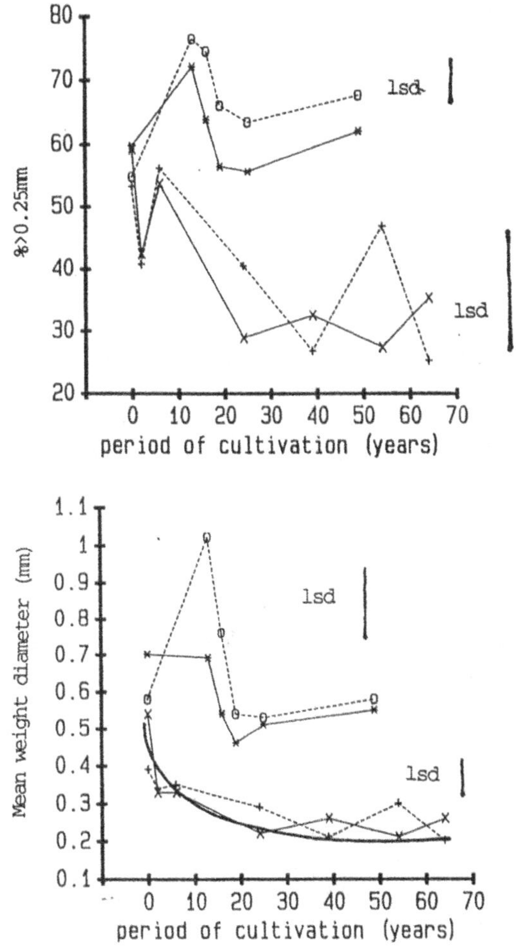

Fig. 3.27 Relationship between water stable aggregates (WSA) >0.25 mm and period of cultivation (**a**) and mean weight diameter (MWD) (**b**) and period of cultivation in a Typic Pellustert: WSA = 30.4 + (55.7 − 30.4) exp (−0.087 year^{-1}); $R^2 = 0.76$. MWD = 0.25 + (0.53 − 0.25) exp (−0.453 year^{-1}); $R^2 = 0.90$. (Source: adapted from Cook 1988)

2017; Hati et al. 2021). For example, water-stable aggregates >20 μm declined almost five times faster in Alfisols than in Vertisols (Figs. 3.27 and 3.28). The half-life, $t_{½}$, of water-stable aggregates >20 μm, in Alfisols was found to be less than 6 years ($t_{½}$= ln 2/0.117) compared to about 28 years ($t_{½}$= ln 2/0.025) in Vertisols. The rates of loss of organic C from these Alfisols and Vertisols under mechanical cultivation and cereal cropping for up to 70 years were inversely related to the percentage of water-stable aggregates >20 μm (Dalal and Mayer 1986b). Thus, organic matter in soil not only stabilises the aggregates (Tisdall and Oades 1982; Field and Minsan 1999; Six et al. 2004) but itself may be protected within aggregates by becoming relatively inaccessible to proteolytic and degradative enzymes and microorganisms. Using ^{13}C NMR techniques, Skjemstad et al. (1986) showed that in these Vertisols, organic matter is protected from loss by oxidation through its intimate association with the soil clay minerals and hence aggregation. In Ultisols

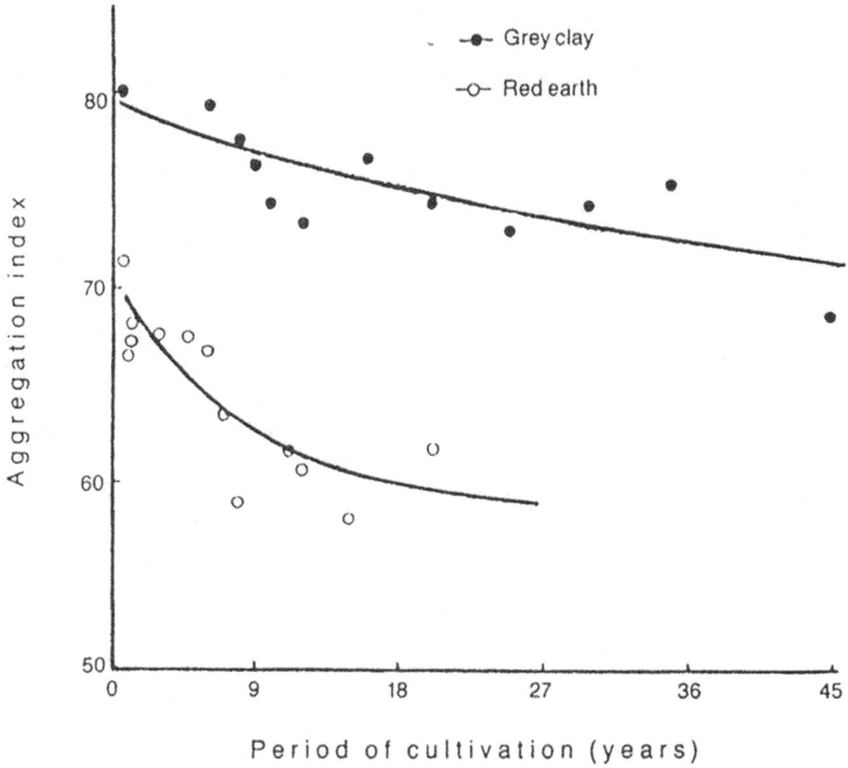

Fig. 3.28 Decline in soil aggregation index (AI) with increasing periods of cultivation in Langlands-Logie (grey clay Vertisol) and Riverview (red earth Alfisol) soils: Alfisol: AI (%) = 58.3 + 11.6 exp (-0.117 year^{-1}), $R^2 = 0.67$, Vertisol: AI (%) = 67.2 + 12.2 exp (-0.025 year^{-1}), $R^2 = 0.54$. (Source: Dalal et al. 1991)

and Oxisols, aggregate stability is also reduced by land clearing and cultivation (Seubert et al. 1977; Moreau 1978; Alegre and Cassell 1986). Legout et al. (2005) showed that the breakdown of aggregates with the aggregate stability tests and under rainfall was similar under both soil types (clay loam and silt loam), and it is apparent that it is qualitatively independent of the size of initial aggregates. Several other researchers have also quantified the aggregate breakdown in different soils and corroborated these findings (Glanville and Smith 1988; Sarmah et al. 1996, Pillai and McGarry 1999).

The severity of mechanical cultivation (also animal treading, mechanical shearing, and earthworks) on the aggregate breakdown is felt most in non-swelling clay soils, especially at low SOM content. In swelling clay soils, such as Vertisols, other aggregate-stabilising agents, clay content, electrolyte concentration, exchangeable sodium, and pH primarily affect aggregate stability (Dalal 1989; Pillai and McGarry 1999). The role of organic matter in aggregate stability in these soils is ambiguous since it is interrelated and affects other aggregate stabilising agents as well (Prebble 1987; Dalal 1989).

Sub-surface aggregate breakdown resulting in compaction and hard-setting, especially in Alfisols and Inceptisols, is much less understood, although SOM loss, mainly macroorganic matter (particulate organic matter > 50 μm), almost invariably leads to soil fertility degradation (Fig. 3.28).

Thomaz et al. (2022) evaluated the aggregate stability of five Oxisols developed from igneous rocks. They inferred that soil aggregate stability behaved according to hierarchical resistance but differed in each breakdown mechanism, namely, *abrasion*, *slaking*, *differential swelling*, and *raindrop impact*. Overall, the larger aggregates (4–8 mm) responded to a lower resistance to abrasion and raindrop impact. However, the smaller aggregates (1–2 mm) were more affected by slaking than raindrop impact. The SOC, except for large aggregates, and soil texture did not show a significant role in aggregate stability. Their findings further suggested that rainfall simulation (raindrop impact) and wet sieving (fast wetting) were the recommended methods to study the (sub)tropical soil aggregate stability. Besides, in tropical and subtropical regions, rainfall volume and erosivity play a critical role in the breaking down of soil aggregates.

Nature of Fertility Degradation

4

Abstract

The nature of fertility degradation includes soil organic matter loss, which leads to the depletion of organic C, N, organic P and organic S. Losses of organic C following land use change from native vegetation (forests, woodlands, grasslands) to cropping vary from 20 % to 80 %, and besides soil fertility degradation, has contributed almost 120 x 1015 g of CO2-C to the atmosphere over the last 12,000 years. Since N, organic P and organic S are closely linked, loss of SOM also leads to the depletion of N, P and S. In addition, SOM loss also leads to the depletion of energy source to microorganisms hence microbial biomass loss, loss of cation exchange capacity in low activity clay soils and coarse-textured soils, reduced pH buffer capacity, reduction in complexed micronutrients, and increased aggregate breakdown due to reduced supply of aggregate forming and stabilising organic agents.

Key words

Soil organic matter · C · N · P · S · Aggregate breakdown · Aeration · Microbial activity · Soil-borne plant pathogens · Greenhouse gases

The characteristic nature of soil fertility degradation includes soil organic matter (SOM) loss, depletion of nutrient elements, decline in nutrient element bioavailability, soil physical degradation, and biological degradation (build-up of soil-borne plant pathogens) (Campbell 2017; Rasool et al. 2007; Mosier et al. 2021). Most of these aspects are interrelated and are caused by one or more of the processes of fertility degradation (Chap. 3).

R. C. Dalal, S. Jayaraman, *Soil Fertility Degradation, and Soil Health Restoration and Management*, https://doi.org/10.1007/978-981-96-0405-0_4

4.1 Organic Matter (SOM) Loss

Organic matter exercises a vital role in most cultivated soils; for example, it is a source of nutrient elements (such as N, P, S) in contributing to the maintenance of a favourable soil structure, in the retention of cations, in complexing micronutrient elements (such as Cu, Zn, Fe, and Mn), and in providing buffer power in low-clay soils. Hence, its rate of loss from the soil following cultivation is an important determinant of soil fertility depletion and degradation (Vityakon 2005; Wiesmeier et al. 2019), and therefore, sustaining SOC (and SOM) stocks plays a significant role in maintaining soil fertility (Tiessen et al. 1994; Dalal and Carter 2019).

One of the primary causes for loss of SOM is the reduced C inputs and increased C outputs from the conversion of natural system (i.e. forest land use/perennial vegetation) to cultivation, which leads to 25–75% loss of the original stock (Guo and Gifford 2002; Kopittke et al. 2016; Dalal et al. 2021; Lal 2018; Lorenz and Lal 2018; FAO 2019; Dubey 2022; Georgiou et al. 2022).

Cultivation of soil that previously supported native vegetation or pasture generally leads to a reduced level of SOM (Clarke and Marshall 1947; Haas et al. 1957; Russell 1981; Odell et al. 1984; Rasmussen and Collins 1991; Six et al. 2000a, 2000b; Guo and Gifford 2002, Dalal and Mayer 1986b; Dalal et al. 2011a, b, c, 2021; Sarker et al. 2018; FAO 2019). The organic matter level of a cultivated soil eventually attains a steady state, where the rate of formation of new SOM from organic residues (plant and crop residues, roots and root exudates, organic wastes, manures, and green manures) equals the rate of SOM decomposition, provided that soil and crop management practices remain essentially similar over a long period (West and Post 2002; West and Marland 2002). Temperature and plant available water (and nutrients) are the major determinants of plant biomass production, and along with cultivation and soil type, they determine the rate of change in SOM as well as SOM stock at a steady state (Paustian et al. 1997, 2016).

4.1.1 Description of Rate of SOM Loss

Since the net concentration or stock of SOM depends upon the relative rates at which organic materials are added to the soil and lost from it through decomposition, it can mathematically be expressed as follows (Jenny 1941, 1994; Bartholomew and Kirkham 1960):

$$\mathrm{d\ SOM/dt = A - k\ SOM} \quad (4.1a)$$

$$\text{or} \quad \mathrm{SOM_t = SOM_0\ \exp\ (-kt) + A/k\ [1 - \exp\ (-kt)]} \quad (4.1b)$$

where SOM_0 and SOM_t are the SOM concentration or stock initially ($t = 0$) and at a given time, t, respectively; a is the rate at which organic matter is returned to the soil, and k is the rate of loss of SOM. Since SOM is heterogeneous in composition, its various components decompose at different rates, and Eq. 4.1b can be modified (Woodruff 1949) to

$$SOM_t = SOM_1 \ \exp\ (-k_1 t) + A_1/k_1\ [1 - \exp\ (-k_1 t)]\ +\ \ldots$$
$$+\ SOM_n \exp\ (-k_n t) + A_n/k_n$$
$$[1 - \exp\ -(k_n t)] \tag{4.1c}$$

where SOM_1, SOM_n refer to different components of organic matter, such as microbial biomass (labile), light fraction or free particulate organic matter (fPOM) (active, humified SOM), aggregate-occluded POM (oPOM), and heavy fraction (passive, inert SOM) or mineral-associated organic matter (MAOM) (Dalal and Mayer 1986c, d, 1987a, b; Parton et al. 1987; Jenkinson and Rayner 1977; Jenkinson and Johnston 1977; Jenkinson and Parry 1989; Jenkinson 1990; Jenkinson et al. 1992; Daroch et al. 2024), and k_1 and k_n refer to the rate of loss of individual components. When $A_t = 0$, at time *t*, Eq. 4.1a is reduced to $SOM_t = SOM_1 + \ldots + SOM_n = SOM_0$ and at time, $t = \propto$, $SOM_t = A_1/k_1 + \ldots + A_n/k_n = SOM_e$, which may be simplified to give $SOM_e = A/k$, where SOM_e is an organic matter content at steady state after a long period of consistently similar soil management (Bartholomew and Kirkham 1960). The simplified form of Eqs. 4.1c and 4.1d is

$$SOM_t = SOM_e + (SOM_o - SOM_e)\ \exp\ (-\,kt) \tag{4.1d}$$

where k is the overall rate of loss of various organic matter components. The turnover period of SOM is then reciprocal of the k value. Equation 4.1d is almost exclusively used here to describe the rate of loss of SOM subjected to continuous cultivation and cereal cropping. Russell (1975) extended Eq. 4.1d to include the feedback effect of crop yield on SOM and SOM dynamic models, which consider not only the effect of organic matter addition but also the environment, soil matrix, cultural practices, and some soil degradation processes such as soil erosion (Jenkinson and Rayner 1977; Van Veen and Paul 1981; Van der Linden et al. 1987; Parton et al. 1987, 1988, 1992; Six and Jastrow 2002).

Of late, various models have been developed for estimating SOM turnover in soil. SOMM, a simulation model of SOM mineralisation, humification, and nitrogen release, has been elaborated by Chertov and Komarov (1997). Falloon and Smith (2000) developed a model for simulations of SOM turnover for the Rothamsted Broadbalk winter wheat experiment using the Rothamsted C model and CENTURY. Mohanty et al. (2020) used APSIM and Jha et al. (2021) RothC model for predicting carbon sequestration potential and saturation from the measured SOC stocks in a 43-year-old long-term fertiliser experiment (LTFE) and three LTFE experiments, respectively, in a number of soil types (Inceptisol, Alfisol, Vertisol) in India.

4.1.2 Turnover of SOM in Natural Ecosystems

The natural ecosystems and grasslands and forests account for almost 90% of SOM (Post et al. 1982; Bai and Cotrufo 2022), although estimates of SOM vary widely (Georgiou et al. 2022). The estimates of SOM vary from a low level of 700×10^{15}g C

(Bolin 1970) to a high of 2946 × 10^{15}g C (Bohn 1976); however, most estimates converge towards a narrow range of 1300–1600 × 10^{15}g C (Meentemeyer et al. 1985; Eswaran et al. 1993). Based on Holdridge life zones (variables: temperature, precipitation (P), and potential evapotranspiration (PET)), the tropical and subtropical regions account for almost 30% of the total SOM. The SOM stocks were closely related to the annual PET/P ratio, where PET/P was 1 and SOM stocks were 10 kg C m^{-3} and decreased as the PET exceeded P and increased as P exceeded PET (Post et al. 1985). Thus, SOM stock is generally lower in semi-arid subtropical regions than in sub-humid subtropical regions. Another estimate reported by Bai and Cotrufo (2022) showed that grasslands possess the third largest global terrestrial C stock and vital source of SOC sink. They reported that the achievable SOC sequestration potential in global grasslands is 2.3 to 7.3 billion Mg of carbon dioxide equivalents per year (CO_2e $year^{-1}$), for biodiversity restoration, and 148 to 699 and 147 million Mg of CO_2e $year^{-1}$, for improved grazing management and sown legumes in pasture lands, respectively.

Organic matter turnover in soil at a steady state can be estimated from Eq. 4.1a, especially in natural ecosystems where it can reasonably be assumed that organic matter additions balance organic matter losses (Table 4.1) unless either or both of these variables are affected by global warming, increasing CO_2 concentration (CO_2 fertilisation), atmospheric N deposition, and changing annual precipitation amounts and distribution. Total amounts of organic C in different zones are the estimates of Post et al. (1982), and annual additions of organic C were calculated using the Rothamsted model (Jenkinson et al. 1991). The Rothamsted model provided about

Table 4.1 Amount of organic C (Ce), annual addition (A), rate of loss (k) and period of turnover (1/k) of organic C in soil under natural ecosystems in warm Holdbridge life zones

Life zone	Total amount[a] (Ce) (×10^{15} g C)	Annual addition[b] (A) (×10^{15} g C $year^{-1}$)	Rate of loss[c] (k) ($year^{-1}$)	Period of turnover (1/k) (year)
Warm desert	20	0.6	0.030	33.0
Tropical desert bush	2	0.1	0.050	20.0
Tropical woodland and savanna (semi-arid)	129	11.5	0.089	11.2
Tropical forest very dry (arid)	22	1.7	0.077	13.0
Tropical forest dry (semi-arid)	24	1.1	0.046	21.7
Tropical forest moist (sub-humid)	60	13.2	0.220	4.5
Tropical forest wet (humid)	78	15.3	0.196	5.1
Total[d]	335	43.5	0.130	7.7
World Total	1394	75.8	0.054	18.5

[a]Post et al. (1982). [b]Jenkinson et al. (1991). [c]Calculated from Eq. 4.1a, assuming organic C is at steady state in natural ecosystems, that is, $dC/dt = 0$, and $A = k\ Ce$ or $k = A/Ce$. [d]Tropical and subtropic wetlands account for additional 100–150 × 10^{15} g C, and thus total organic C in soils of world in the top 1 m is about 1550 × 10^{15}g (Eswaran et al. 1993)

10–15% higher estimates of net primary production (that is, organic C input) than earlier estimates (Buringh 1980, 1984; Ross 1993), although Long et al. (1989) measured two to five times higher net primary production (NPP) in natural grass ecosystems of the tropics.

The calculated rates of loss of SOC (or SOM) vary from 0.03 $year^{-1}$ in arid deserts to 0.2 $year^{-1}$ in sub-humid and humid tropical forests, although in semi-arid tropical forests, the rate of loss of organic C is only 25% of that in the sub-humid tropical forests and conversion of natural forest (tropical rain forest to rubber plantation led to 50% loss of SOC in Aric Ferralsols, in Yunan Province, Southern China (Balasubramanian et al. 2020; Li et al. 2022; Zhang et al. 2024). The rate of SOM loss in semi-arid woodlands and savannas in the tropics (and subtropics) is less than 0.1 $year^{-1}$, a value between dry tropical forests and wet tropical forests. Across the different zones, Wei et al. (2014) analysed 453 data sets of converted natural forest to agriculture and found the largest decrease in SOC stocks of 52% (in temperate regions), followed by a 41% decrease (in tropical regions), and a 31% decrease (in boreal regions). Climate and soil factors greatly affected the decrease in SOC stocks. Moreover, the percent decrease of SOC stocks and turnover rate varied drastically depending upon forest type and cultivation stage.

The period of turnover of SOM is about 2.5 times faster in the tropical and subtropical regions than that for the whole terrestrial ecosystem (Table 4.1), and in the tropical region alone, it is about four times faster than that in the temperate region (Jenkinson and Ayanaba 1977). In the sub-humid tropical forest, the rate of turnover of SOM is less than 5 years. It is more than two times faster than in semi-arid tropical woodlands and grasslands and four times faster than in semi-arid tropical forests, a reflection of favourable temperature and moisture, both for plant biomass production as well as decomposition.

4.1.3 SOM Loss from Cultivated Soils

4.1.3.1 Overall SOM Loss

The rates of loss and turnover period of SOM in different soils vary considerably (Table 4.2). Organic matter from Ultisols/Alfisols turns over much faster than Inceptisols and Vertisols. At least, part of the effect is due to decomposition being more rapid under the warmer and moister conditions existing in the sub-humid Ultisols/Alfisols.

The rates of loss of SOM from the cultivated tropical and subtropical soils are generally within the range of those of natural ecosystems in these regions (Tables 4.1 and 4.2) except in Alfisols and Ultisols which have much higher rates of SOM loss, especially within the first 8–20 years after land clearing. Conversely, there is a greater possibility of C storage/sequestration through best management practices in these soils (Batjes and Sombroek 1997). Jaiarree et al. (2011) studied the effect of conversion from forest to maize cultivation, which reduced SOC stocks at the rate of 6.97 Mg C ha^{-1} $year^{-1}$ (for the duration of 12 years). This reduction was most prominent in the top 10 cm soil layer (which translates to about 47% loss of SOC

Table 4.2 Rates of organic matter loss and its turnover period from cultivated soils

Location	Soil	Cropping system	Rate of loss ($year^{-1}$)	Turnover period (year)
Illinois, USA[a]	Inceptisol	Continuous corn	0.024	41.7
Adelaide, Australia[b]	Alfisol	Wheat	0.025	40.0
Narayen, Qld, Australia[b]	Vertisol	Sorghum	0.050	20.0
Southern Qld, Australia[c]	Vertisol	Wheat	0.041–0.178	5.6–24.4
Kansas, USA[d]	Argiustoll	Wheat-fallow	0.062	16.1
Assam, India[e]	Ultisol	Unshaded tea	0.099	10.1
Congo, Africa[a]	Ultisol/Alfisol	Food crops	0.330	3.0
Nigeria, Africa[f]	Ultisol	Food crops	0.393	2.5
Nigeria, Africa[g]	Alfisol	Maize	0.637	1.6
Southern Qld, Australia[c]	Alfisol	Wheat	1.211	0.8
World cultivated lands[h]	All soils	All crops	0.061	16.4

[a]Adapted from Lathwell and Bouldin (1981); [b]Russell (1981); [c]Dalal and Mayer (1986b); [d]Hobbs and Thompson (1971); [e]Gokhale (1959); [f]Calculated from Obi (1989); [g]Calculated from Lal and Kang (1982); [h]Calculated from Jenkinson et al. (1991)

after 12 years of cultivation). They reported that low input rates of maize-derived C were not sufficient to maintain SOC at the original level. Moreover, the maize-derived C fraction contributed only 20% of total SOC (i.e. 5 Mg ha^{-1} of the total 25.31 Mg ha^{-1}).

Wang et al. (1999) studied paired natural forest and disturbed soils and found that clearing and continuous cultivation of soils led to a reduction in SOC. The loss in soil C stock varied between 26% and 30%, with more SOC lost from the upper horizons. They reported that land use change could have a significant impact on the soil C cycle. The average turnover times of organic matter are longer in disturbed soils than in natural soils due to preferential loss of C (i.e. loss from "active" soil C pools). In both natural and disturbed soils, the mean turnover times of SOM increase with depth from decades or less in shallow soil layers to hundreds of years or even thousands of years in deeper soil layers.

Dalal et al. (2011b) reported that land use change from C_3 vegetation ($\delta^{13}C$ values, –30‰ to –24‰) to C4 vegetation ($\delta^{13}C$ values, –14‰ to –11‰) provides a useful quantitative technique for estimating SOC turnover in soil, even when total SOC changes are negligible. They physically fractionated SOM into POM C (>250 μm fraction and POM C (250–53 μm fraction) and the <53 μm mineral-associated organic matter (MAOM C) fraction. The SOC turnover periods (years) were 31 ± 6 for the 0–0.1 m depth and increasing with depth: 60 ± 5, 55 ± 15, and 63 ± 20 for the 0.1–0.2, 0.2–0.3, and 0.3–0.4 m depths, respectively; the corresponding turnover periods for the POM C >250 μm fraction were 13 ± 2, 19 ± 5, 14 ± 4, and 12 ± 5 years, respectively. The turnover periods of SOC in the POM C 250–53 μm and <53 μm fractions were similar to, or longer than, the whole SOC at all depths studied. Thus, the lability of the SOC and SOC pools was in the order: POM C >250 μm fraction > POM C 53–250 μm fraction = MAOM C <53 μm fraction.

In a number of these studies reported in Table 4.2, the effect of change in bulk density with the loss in SOM on the depth of sampling is not considered; hence, equivalent sampling depths (or equivalent soil mass) may differ from actual sampling depths, so these estimates of rates of SOM loss should be considered approximate and should be corrected for the long-term monitoring of SOC stocks.

These considerations were taken into account in SOM studies by Dalal and Mayer (1986a, b, c, d) who investigated the long-term trends in fertility of Vertisols and Alfisols under continuous cultivation and cereal cropping in the semi-arid subtropical region (southern Queensland, Australia). Six major soils (five Vertisols and one Alfisol) of this cereal belt, cropped for maximum periods of 20–70 years, were examined. These were Waco clay soil (Typic Pellusterts, 72% clay), Langlands-Logie clay soil (Typic Chromusterts, 49% clay), Cecilvale clay soil (Typic Chromusterts, 40% clay), Billa loamy clay soil (Typic Chromusterts, 34% clay), Thallon clay soil (Typic Chromusterts, 59% clay), and Riverview sandy loam soil (Rhodic Paleustalfs, 18% clay). The ranges of mean annual rainfall and temperature (83–101 years period) for the areas concerned were 480 mm and 670 mm and 18.5 ° C and 20.5 °C, respectively. Winter crops were wheat (90%), barley (9%), and oats (1%) and summer crops were sorghum (90%) and sunflower (8%). The use of fertilisers was restricted to only a small, eastern area of the study (Dalal and Mayer 1986a).

The changes in SOM with different periods of cultivation (tillage) and cereal cropping in the six soils from southern Queensland (148–152 °E and 26–29 °S) are shown in Fig. 4.1 (Dalal and Mayer 1986b). All soils lost SOC when cultivated, and the loss of SOC increased with increasing periods of cultivation. The soils differed,

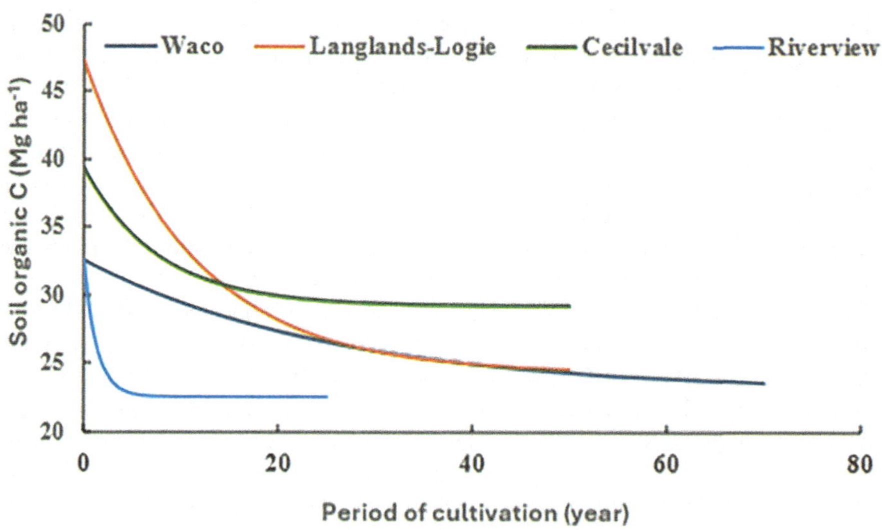

Fig. 4.1 Decrease in organic C stock in the top layer (0–0.1 m) with a period of cultivation. Curves are drawn according to Eq. 4.1d for each soil. Waco, Langlands-Logie, Cecilvale, Billa Billa, Thallon, Riverview. The rate of organic C loss (k) is the least from Waco soil (Typic Pellusterts) and the most from Riverview (Rhodic Paleustalfs). (Source: Dalal and Mayer, 1986b)

Table 4.3 Initial values, steady-state values, overall rate of loss, rate of addition and period of turnover of organic C in six soil series (0–0.1 m depth)[a]

Soil series	Soil taxonomy	Initial C Co (Mg C ha^{-1})	Steady-state C Ce (Mg C ha^{-1})	Rate of loss k ($year^{-1}$)	Rate of addition[b] A (Mg C ha^{-1} $year^{-1}$)	Period of turnover (1/k) (year)
Waco clay	Typic Pellusterts	13.71	8.26	0.065	0.54	15.4
Thallon clay	Typic Chromusterts	7.52	4.41	0.069	0.30	14.5
Langlands-Logie clay	Typic Chromusterts	22.07	7.76	0.080	0.62	12.5
Cecilvale clay	Typic Chromusterts	17.51	10.18	0.180	1.83	5.6
Billa Billa loamy clay	Typic Chromusterts	13.64	8.27	0.259	2.14	3.9
Riverview sandy loam	Rhodic Paleustaffs	15.69	9.38	1.224	11.48	0.8

[a]Dalal and Mayer (1986b). [b]Calculated from Dalal and Mayer (1986b)

however, in the rates of loss, rates of addition, and periods of turnover of SOC (Table 4.3).

Dalal and Mayer (1986b) observed that, in the contiguous region, initial SOC concentrations (virgin soils) as well as SOC concentrations after 20 years of cultivation for cereal cropping were closely correlated with the mean annual precipitation. However, the rate of increase in SOC concentrations for each mm of precipitation was almost twice that of SOC concentrations in the cultivated soils (4.8×10^{-3} SOC % mm^{-1} vs. 2.9×10^{-3} SOC% mm^{-1}). This was primarily due to the lower amount of organic residues not only produced but also returned to the soil and/or accelerated SOM decomposition in cultivated soils due to the aggregate disruption and increased decomposition of exposed SOM. Increased incidence of erosion may also be a contributory factor (Van Veen and Paul 1981; Cook et al. 1992). This is supported by the observation that in wider regions, however, the P/PET ratio determines the plant biomass production (Post et al. 1985; Parton et al. 1988).

In the six-soil series studied, Dalal and Mayer (1986b) recorded a 20-fold difference in the overall rate of SOC loss and the period of turnover of SOM (Table 4.3). The rate of SOC loss declined exponentially as the clay content in soil increased (Fig. 4.2), irrespective of clay mineralogical composition. Similarly, the overall rate of loss of SOC declined as aggregation in soil increased (Jayaraman et al. 2017, 2018a, b, c, d, 2024), thus reducing the accessibility of SOC to microorganisms, degradative enzymes, and other SOM decomposers. It is, therefore, primarily the clay that protects SOM from decomposition when soil is cultivated (Sorenson 1981; Dalal and Mayer 1986b; Skjemstad et al. 1986; Skjemstad and Dalal 1987; Parton et al. 1988; Six et al. 2000a, b; Dalal et al. 2011a; Six and Paustian 2014; Jayaraman et al. 2017; Sarker et al. 2018; Jha et al. 2020; Hati et al. 2021). Using spectroscopic techniques, including ^{13}C nuclear magnetic resonance

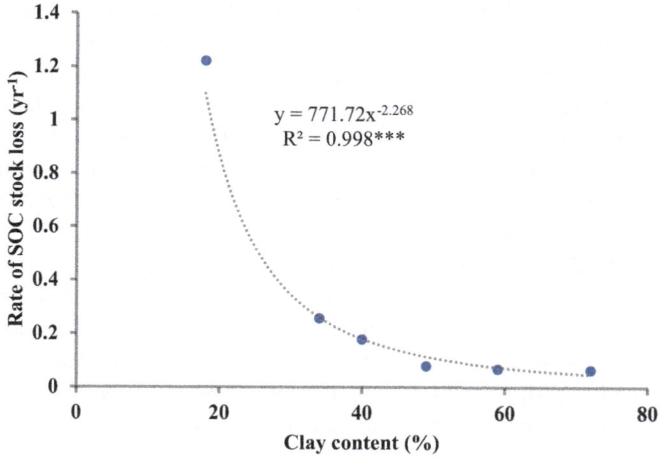

Fig. 4.2 Rate of SOC stock loss (Kw) as a function of the clay content of six-soil series. (Drawn from the data of Dalal and Mayer 1986b)

Table 4.4 Effect of cultivation on soil organic C distribution in Langlands-Logie soil (Typic Chromusterts) using solid state CP/MAS ^{13} C-NMR technique

Cultivation (year)	Organic C (%)	C distribution (%)[a]			Carbohydrate-C[b]
		Polymethylene	Methoxyl and carbohydrate	Aromatic and carboxyl	Total organic-C
0	2.30	65	31	4	5.4
25	1.14	63	25	12	–
35	0.89	62	22	16	8.9
45	0.77	52	39	9	9.6

[a]From Skjemstad et al. (1986)
[b]In soil acid hydrolysates using chromatographic technique (Dalal and Henry 1988)

spectroscopy (Table 4.4), Skjemstad et al. (1986) showed that the chemical nature of SOM remained essentially similar in spite of 60% SOC lost upon cropping, in that poly-methylene and methoxyl-C were the dominant groups of organic C throughout the SOC loss process. Similar results have been obtained by others (Oades 1988; Arshad et al. 1990, 2016; Capriel et al. 1992). Skjemstad and Dalal (1987) also observed that besides the association of SOM with clay, SOM stability against decomposition in cultivated soils may be due to increased aromaticity as well as shorter, more highly branched alkyl chains (pyrogenic C), at least in some Vertisols (Pellusterts).

Dalal and Mayer (1986b) also observed that the rate of organic C loss in the six-soil series decreased as the ratio of organic C/urease activity increased, which is a measure of relative accessibility of organic C to proteolytic and other degradative enzymes (Roscoe et al. 2000). Pettit et al. (1976) postulated that extracellular urease may be stabilised in soil by bonding or entrapment in the organic matter matrix which renders it inaccessible to attack by soil proteolytic enzymes. The ratio of organic C/urease activity as an index of accessibility to or protection of organic matter needs to be confirmed in different environments. Further, the rate of SOC loss was negatively related to the aggregation index (Dalal and Mayer 1986b). This concept is somewhat akin to that of soil micropores small enough to be inaccessible to microorganisms (McGill and Myers 1987) and proteolytic enzymes; hence, SOM in that environment was protected from microbial decomposition.

The equations that have been used above to describe the trends in soil organic carbon under continuous cultivation are also helpful in understanding why it is so difficult to restore soil SOM if SOM stocks have been allowed to decline. There is a relation between the steady-state content of SOM and the annual rate of addition of SOM to the soil ($SOM_e = A/k$); that is, the rate of input, A, equals the product of SOM_e and k output under constant soil and crop management practices (Tables 4.2 and 4.3).

It has been shown that the rate of decomposition (k) is greater in coarse-textured soils, and it is also higher under warm and moist conditions. To maintain a given steady-state concentration and stock of SOM, inputs must be higher on soils that have a higher rate of decomposition (Six et al. 2002; Dalal et al. 2011a, b; Conrad et al. 2018). Further, the necessary increase in annual inputs to maintain the SOM concentration and stock at a higher steady state is also greater where the rate of decomposition is high. The concentration and stock of SOM that can be maintained ultimately depend on the production of plant biomass and/or the addition of exogenous C such as farmyard manure and organic wastes. It is apparent, therefore, that the SOC declines under continuous cropping for at least two reasons: rates of decomposition are increased due to cultivation and annual inputs are lower under a cropping system than they were under a grassland or forest ecosystem (Dalal et al. 2011a, b; Six and Paustian 2014). In addition, biomass C in the produce is removed from the site. Moreover, soil aggregates are disrupted, thereby exposing new SOM, which otherwise was relatively inaccessible to microbial decomposition.

4.1.3.2 Loss of SOM from Various Soil Fractions

As mentioned above, a more complete understanding of the dynamics of SOM requires knowledge of the various components of organic C and their rates of decomposition (Eq. 4.1c) (Coleman and Jenkinson 1996). Jenkinson and Rayner (1977) proposed five components of SOM with different turnover periods, namely, decomposable plant material, resistant plant material, soil microbial biomass, and physically and chemically stabilised SOM, with turnover periods of 0.25, 3.33, 2.44, 71.4, and 2857 years, respectively. van Veen and Paul (1981) incorporated modifications of this model that accounted for the protection of OM by soil constituents, primarily clay. Under grassland, it was assumed that 50% of the SOM was protected, whereas it was only 20% protected under cultivation.

Parton et al. (1987, 1992a, b) extended the concepts of Jenkinson and Rayner (1977) and van Veen and Paul (1981) into their 'CENTURY' model for simulating the dynamics of C, N, P, and S in different types of ecosystems (grasslands, forest,

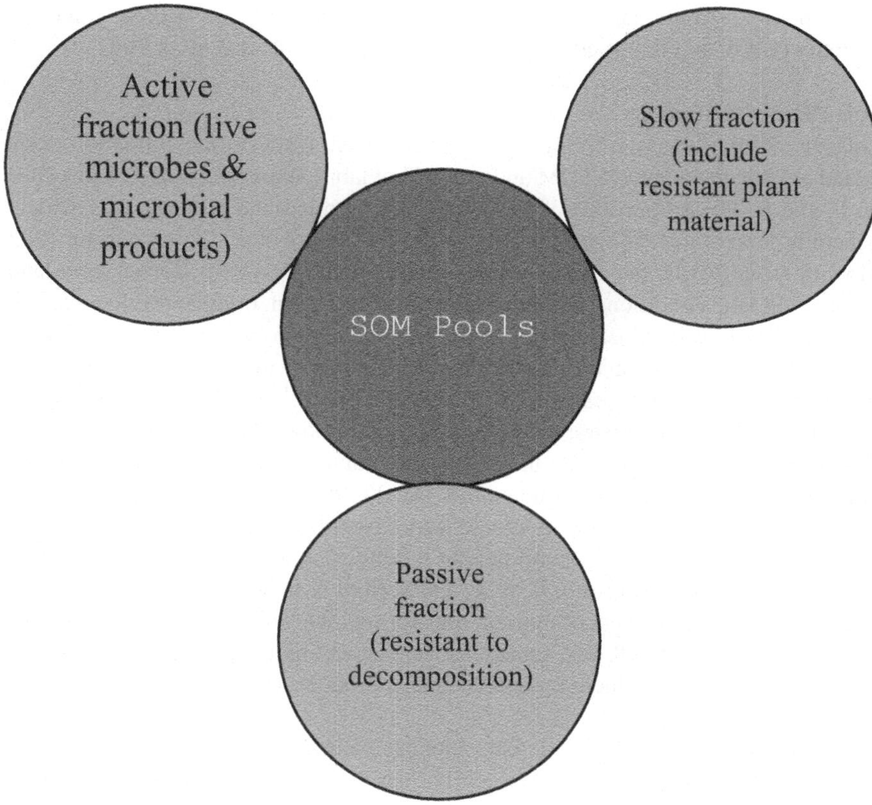

Fig. 4.3 Different SOM fractions/pools used in the carbon models

crops, and savannas). The SOM components of their model constitute a '*surface live microbial pool*', an '*active*' fraction (live microbes and microbial products), a '*slow*' fraction which is considered to include resistant plant material and soil-stabilised plant and microbial material, and a '*passive*' fraction that is very resistant to decomposition and includes physically and chemically stabilised SOM. These last three fractions typically are assumed to have turnover times, for a grassland site, of 2, 40, and 2000 years. Unfortunately, the active, slow, and passive fractions of SOM in the CENTURY model are only conceptual components of the SOM (Fig. 4.3). There is a need to experimentally determine appropriate portions of the total SOC and their turnover rates in cultivated soils. In addition to the carbon model, this model also comprises a water budget model, a grassland/crop submodel, a forest production submodel, and management and event scheduling functions (Metherall et al. 1993).

The Roth C model is also able to predict the turnover of SOM in non-waterlogged soils from years to centuries (Coleman and Jenkinson 1996; Coleman et al. 1997; Jha et al. 2021). This model comprises four active compartments and a small amount of inert organic matter (IOM). Similarly, the four active compartments include decomposable plant material (DPM), resistant plant material (RPM), microbial biomass (BIO), and humified organic matter (HUM) (Coleman et al. 1997).

Soil Microbial Biomass

Soil microbial biomass, a living part of SOM, is an agent of transformation of the added organic residues and SOM and a small but labile source, as well as a sink of C, N, P, and S. The proportions of SOC contained in microbial biomass vary widely, generally, from 1 to 5% (Insam and Domsch 1988; Insam et al. 1989; Sparling 1992; Silveira et al. 2014), depending on the nature of an ecosystem (forest, grassland, arable), climatic conditions, seasons, cropping, cultivation, organic residue management, and fertilisation. Insam et al. (1989) found that microbial biomass C (MBC) as a proportion of total organic C decreased exponentially as the ratio of precipitation to pan evaporation (P/E) increased [MBC/organic C (g kg^{-1}) = 18.18 + 108.87 exp {− 6.7 P/E (mm)}]. In a geographically contiguous region, Dalal and Mayer (1987b) found that MBC increased with increasing mean annual precipitation and decreased with increasing mean annual temperature. Dalal (1998a, b) reported that soil microbial biomass comprises < 5% of organic matter in soil, and it is considered a sensible indicator to assess the state of soil health. It delivers at least three critical functions in soil: (i) it is a labile source of C, N, P, and S; (ii) it is an immediate sink of C, N, P, and S; and (iii) it is an agent of nutrient transformation/cycling, as well as facilitating pesticide degradation. Of late, molecular genetic techniques help in understanding the role of biotic factors including MBC and its products in aggregation processes in soil (Kaur et al. 2022).

Continuous cultivation for cereal cropping of originally virgin soils generally decreases microbial biomass more rapidly than SOM (Table 4.5). Therefore, it has

Table 4.5 Rates of loss of organic C from various C fractions of 0 to 0.1 m layer of Waco soil series (Typic Pellusterts)[a]

Organic C fractions	Initial C Co (kg ha^{-1})	Steady state C Ce	Proportional Rate of Loss (%) (Co–Ce)/Co	($year^{-1}$)	Turnover period (year)
Whole soil	13 710	8 260	40	0.065	15.4
Microbial biomass	0 380	0 130	66	0.3382	2.9
Light fraction, <2 Mg m^{-3}	2 860	1 000	65	0.365	2.7
Heavy fraction, >2 Mg m^{-3}	11 080	6 560	40	0.033	30.0
Sand-size fraction, 0.002–2 mm	2 730	0 860	69	0.096	10.4
Silt-size fraction, 0.002–0.02 mm	3 990	2 030	49	0.039	25.6
Clay-size fraction, < 0.002 mm	7 190	5 040	29	0.057	17.5

[a]After 70 years of continuous cultivation and cereal cropping. From Dalal and Mayer (1986b, 1986c, 1986d, 1987b)

been suggested that microbial biomass C is a sensitive indicator of the effects of management practices over relatively short periods compared to SOM (Powlson et al. 1987; Insam et al. 1989; Sparling 1992; Dalal 1998a, b). Returns of reduced amounts of organic residues to cultivated soils would rapidly decrease the amount of MBC. Since MBC is closely related to a suite of biological and biochemical activities in soil (Dalal and Mayer 1987b; Dalal 1998a, b; Dalal 2021; Daroch et al. 2024), the reduced amount of microbial biomass upon cultivation of previously fertile soils leads to soil fertility depletion in three ways: reduced source of labile N, P, and S; reduced retentive capacity (immobilisation) for mineral N in most soils and, possibly, P and S in plant available form, in Ultisols, Oxisols, and Alfisols; and reduced rate of mineralisation and thus lower bioavailability of nutrients (Sect. 4.3). Furthermore, continuous cultivation for cereal cropping, especially monoculture, not only reduces the total amount but also alters the composition of microbial biomass and often leads to biological degradation (increased incidence of soil-borne pathogens and reduced disease suppressiveness) (Dalal et al. 2011a; Dalal 2021).

Light Fraction, <2 Mg m^{-3} ('Active' Fraction, free particulate organic matter) The light fraction (<2 Mg m^{-3}) contains most of the partly decomposed plant residues, microflora, and microfauna debris (Ford et al. 1969) and some microbial biomass such as fungal hyphae (Oades and Ladd 1977, Oades 1988; Dalal 1998a, b; Wang et al. 2004). It contains about 25% of SOM, although it varies from 1% to 85% (Greenland and Ford 1964; Christensen 1992), depending on the nature of the ecosystem, vegetation, climate, and soil characteristics. Accumulation of light fraction C in soil may occur in cooler and drier climates, low pH, and forest and grassland ecosystems with high return of plant residues (Wang et al. 2004; Dalal et al. 2005; Silveira et al. 2014).

The rate of loss of light fraction C in cultivated soils is very rapid and essentially similar to that of MBC (Table 4.5) unless this fraction also contains charcoal C originating from anthropogenic burning of vegetation (Skjemstad et al. 1990). In five fine-textured (>30% clay) soil series, the rate of loss of light fraction C upon cultivation was 2–11 times faster than that from heavy fraction (<2 Mg m^{-3}), although a similar rate of C loss from both fractions was observed in a coarse-textured soil (Dalal and Mayer 1986d). Rates of SOM loss in these soils have been associated with soil aggregation and clay content which provide protection against SOM decomposition by microorganisms and proteolytic and degradative enzymes (Dalal and Mayer 1986b; Dalal 1998a, b). The coarse-textured soil, therefore, may provide less protection even for the heavy fraction of organic C which is apparently associated with the inorganic minerals.

Upon cultivation, more than 60% of the light fraction SOM can be lost from the soil within 20 to 70 years (Table 4.4) (Dalal and Mayer 1986a, b, c, d, e; Skjemstad and Dalal 1987; Dalal 1998a, b; Dalal 2021), and similar to the microbial biomass, the proportion of light fraction to total SOM provides an earlier and a sensitive indication of the consequences of different soil management practices than the level of total SOM (Christensen 1992). Moreover, since crop residues and other organic materials in soil maintain active microbial biomass as well as light fraction, non-return of these materials deprives soil of its labile and active fraction of organic C (Dalal 1998a, b; Ferraz de Almeida et al. 2019; Dhillon and Van Rees 2017).

Heavy Fraction <2 Mg m^{-3}, ('Slow' or Soil-Mineral-Associated Organic Matter)

About 75% and 90% of SOM reside in the heavy fraction of virgin and cultivated soil, respectively (Dalal and Mayer 1986d; Christensen 1992).

The turnover of SOM from the heavy fraction is much slower (30 years) than that from the light fraction (2.9 years) (Table 4.5). In fine-textured soils, especially, SOM in the heavy fraction is afforded greater protection than that in the light fraction. The ratio of the rate of loss of light fraction C/heavy fraction C generally increases with the clay content, indicating a clearer separation between the relatively accessible light fraction and relatively protected heavy fraction SOM (Tan et al. 2007; Plaza et al. 2019; Zahedifard et al. 2024). Corroboration of these results was provided by ^{13}C nuclear magnetic spectroscopic studies of SOM fractions from the virgin and cultivated Waco and Langlands-Logie soil series (Skjemstad et al. 1986; Skjemstad and Dalal 1987) and organic C dynamics in various particle-size fractions (Dalal and Mayer 1986c). These studies showed that, although the chemical nature of SOM in different fractions remained essentially unchanged, the heavy fraction SOM and clay-size SOM declined much less than the light fraction SOM upon cultivation. Young and Spycher (1979) proposed that light fraction SOM is located at the surface of soil aggregates and is, therefore, readily accessible to the soil organisms which are confined to aggregate surfaces and extra-aggregate space. The heavy fraction SOM, however, is contained in intra-aggregate regions and thus relatively better protected from soil organisms (Wang et al. 2004). This concept is supported by Elliott (1986) who found that the disintegration of macro-aggregates released readily decomposable SOM (Dalal 1998a, b; Six et al. 2000a, b; Six and Paustian 2014).

Similar to SOM, heavy fraction SOM increases with mean annual precipitation, soil clay content, and crop residue retention and decreases with an increase in mean annual temperature, period of cultivation, and crop residue removal (Dalal and Mayer 1986b). In the medium to long term, heavy fraction SOM is a major source of mineral N in soil.

Sand-, Silt-, and Clay-Size SOM

About 50% of SOM is found in the clay-size fraction (Tiessen and Stewart 1983; Christensen 1985; Dalal and Mayer 1986c; Balesdent et al. 1988; Bonde et al. 1992), and the remainder is distributed between sand-size and silt-size fractions. The proportion of SOM between these two fractions depends upon land use, climate, season, silt and clay content, and cultivation (Schweizer et al. 2019). Virto et al. (2008) studied the relationship between soil aggregation and SOC dynamics. They reported that slaking-resistant silt-size aggregates stored 8% SOC as organic matter, and clay minerals stored 28% SOC and as occluded organic particles (8% SOC) in Eutric Cambisol in France. They concluded that stable microaggregates are formed within the silt-size fraction and store organic C both by adsorption and by entrapment of small organic particles.

Generally, SOM in a particle-size fraction decreases with the increase in that particle-size content in the soil, although the clay fraction contains a greater proportion of SOM over a wide range of clay contents. Similarly, soils high in silt content may also contain a high proportion of SOM in that fraction (McKeague 1971), especially if the silt-size fraction also contains micro-aggregates of clay when soil is not completely dispersed (Christensen 1992; Yan et al. 2022).

It has been observed that the SOM enrichment of clay-size and silt-size fractions is inversely related to the proportion of these particles in the whole soil (McKeague 1971; Christensen 1992; R.C. Dalal, unpublished data). For example, the following relationship was found between C enrichment of the clay fraction and clay content:

clay size C/total C = 0.27 + 51.88/% clay, $r^2 = 0.98$; $n = 94$, for clay contents varying from 3% to 84% (calculated from Christensen 1992 and R.C. Dalal, unpublished data). This significance of SOM enrichment of clay fraction and its selective removal in soil erosion on soil fertility depletion has been described in Sect. 3.1.

The sand-size SOM declines rapidly upon cultivation (Table 4.5), mostly due to oxidation, but also due to disintegration to enter silt-size or clay-size fractions (Ladd et al. 1977). The SOM is not associated with silt-size and clay-size mineral particles; the sand-size SOM, which is essentially the light-fraction SOM (Table 4.5; Dalal and Mayer 1986d; Christensen 1992); and microbial biomass also constitute the 'labile' SOM (Dalal et al. 1991). Therefore, the rapid and extensive loss of the 'labile' SOM upon cultivation may result in greater depletion in soil fertility than is apparent from a comparison of total SOM contents of virgin and cultivated soils (Dalal 1998a, b; Dalal 2021; Ferraz de Almeida et al. 2019; Dhillon and Van Rees 2019).

The rate of SOM loss from the silt-size fraction is much slower than from the sand-size and light fraction (Table 4.5). Dalal and Mayer (1986c) observed that the proportion of SOM-C in the silt-size fraction of five Vertisols and one Alfisol remained essentially similar (26–27%) upon cultivation (20–70 years) of virgin soils, although in the sand-size fraction it decreased from 26% to 12% overall, and proportionately, the clay-size fraction SOM-C increased from 48% to 61%.

The rate of loss of SOM-C from the clay fraction is generally lower than that from the other fractions and the whole soil (Table 4.5) (Dalal and Mayer 1986b, c, d; Parton et al. 1992a, b), resulting in an increasing proportion of total SOM-C (>70%) remaining in the clay fraction with cultivation. The clay fraction, therefore, provides protection to SOM or makes it relatively inaccessible against microbial and enzymic attack, since clay-size organic matter separated by soil dispersion mineralises its C and especially N much faster than from the other size fractions (Elliott 1986; Christensen 1992).

'Passive', 'Inert', or 'Stable' SOM Using ^{14}C Dating and ^{13}C Natural Abundance
The overall turnover period of SOM and its various fractions in cultivated soils rarely exceeds 50 years (Tables 4.1, 4.2, 4.3, and 4.5). However, the passive, inert, or stable SOM may have turnover periods ranging from several hundred to a few thousand years (Martel and Paul 1974; Jenkinson and Rayner 1977; Stout et al. 1981; Guillet 1982; Parton et al. 1987; Balesdent et al. 1988; Hsieh 1992; Six et al.

2000a, b, 2002; Six and Paustian 2014; Conrad et al. 2017, 2018; Collins et al. 2019).

Using the ^{14}C dating technique, the turnover period of SOM has been estimated, variously, to be from 250–710 years (Martel and Paul 1974), 850–2970 years (Jenkinson and Rayner 1977; Anderson and Paul 1984; Hsieh 1992), and up to about 5700 years or even longer (Stout et al. 1981). Another study conducted by Paul et al. (1997) suggested that cultivation has resulted in the decline of the C contents but increased the ^{14}C age by an average of 900 year at the surface layer. However, in the sub-surface layer (i.e. 10–20 cm depth), ^{14}C age was 1200 year or older under cultivated and native sites. At 90 to 120 cm depth, a cultivated site SOC with ^{14}C age 7015 year before present (BP) was reported, which was 6000 year older than the surface layer. However, the ^{14}C ages were not correlated to latitude, although it was strongly influenced by the total SOM-C content.

These estimates are based on a number of assumptions inherent in the ^{14}C dating techniques. The two basic assumptions, the constancy in the ^{14}C activity of the atmosphere during the last few thousand years, and minimum fractionation of ^{14}C during physical mixing and chemical and biological transformations leading to the stable SOM are rarely met. It is known that the ^{14}C specific activity of the atmosphere has varied quite considerably over the last 7000 years, and from 1890 to 1954, the ^{14}C specific activity of the atmosphere gradually depleted due to the burning of almost ^{14}C-free fossil fuels, the so-called '*Suess effect*' (Keeling 1979). From 1954 to 1964, it increased by 60%, and since then, it has gradually been declining (Stout et al. 1981). The physical, chemical, and biological processes up to the formation of the stable SOM involve isotopic discrimination (Stout et al. 1981). The other assumptions involved in ^{14}C dating and likely errors from charcoal to estimate the turnover period of the stable SOM have been discussed by Jenkinson et al. (1991), Nate Hoffer and Fry (1988), and Hsieh (1992). Similar assumptions are involved in using ^{13}C natural abundance as a tracer (Balesdent et al. 1990; Mariotti and Balesdent 1990); besides, it is restricted to SOM studies where complete changeover from either C_3 to C_4 vegetation or vice versa has occurred.

Using the ^{13}C natural abundance technique, Balesdent et al. (1988) measured the turnover period of the stable SOM (about 50% of total SOM) in a temperate Alfisol to be 600 to 1400 years. Skjemstad et al. (1990) measured much shorter turnover periods of micro-aggregates, <0.2 mm (relatively stable SOM) in a subtropical Oxisol, in the range of 75–348 years. In a tropical Oxisol under sugarcane for 50 years, the turnover period of SOM from native vegetation in silt and clay fractions appears to be less than 150 years (137–144 years, calculated from Vitorello et al. 1989), although not all SOM in these fractions was from stable SOM (Dalal and Mayer 1986c). Similarly, Martin et al. (1990) inferred, from the SOM turnover in a tropical Oxisol, that stable SOM (having turnover periods of at least several hundred years) forms only a negligible proportion of the total SOM in tropical soils. Skjemstad and Dalal (1987) found that apparently stable SOM in <2.4 Mg m^{-3} fraction in two subtropical Vertisols was less than 7% of the total SOM.

Notwithstanding the limitations of ^{14}C dating and ^{13}C natural abundance techniques and the difficulty in isolating stable C fraction from SOM, the turnover period, as well as the proportion of even the 'passive', 'inert', or 'stable' SOM, would decrease from the temperate to the tropics (Dalal et al. 2011a; Page et al. 2013a, 2013b, 2020a, 2020b). Therefore, the postulated periods of 1000–2000 years in the 'CENTURY' model (Parton et al. 1987, 1989) and 2871 years in the 'Rothamsted' model (Jenkinson and Rayner 1977; Jenkinson et al. 1992) for the stable SOM need to be modified considerably to successfully model the SOM dynamics in the subtropical and tropical regions (Jenkinson et al. 1992; Chilcott et al. 2007).

4.1.4 Modelling SOM Turnover

Modelling SOM dynamics in an ecosystem integrates physical, chemical, and biological processes involved in SOM turnover, and therefore, it may assist in developing a better understanding of SOM in biogeochemical and atmospheric environments. They are useful in formulating and testing hypotheses and in establishing the relative importance of parameters (Verberne et al. 1990; Chertov and Komarov 1997; Paustian 2001; Chilcott et al. 2007).

Most SOM models assume that different components of SOM differ in their rates of decomposition and turnover periods (Jenkinson and Rayner 1977; van Veen and Paul 1981; McGill et al. 1981; Parton et al. 1987, 1992a, b; Verberne et al. 1990; Li et al. 1992; Chertov and Komarov 1997; Paustian 2001;), although most of these are only conceptual components, and therefore, currently, not all the conceptual fractions of SOM have been experimentally verified (Parton et al. 1996).

Jenkinson and Rayner (1977) and Jenkinson et al. (1991) proposed five compartments of organic matter in soil with different turnover periods, that is, decomposable plant material, resistant plant material, soil microbial biomass, and physically and chemically stabilised organic matter with turnover periods of 0.25, 3.33, 2.44, 71.4, and 2857 years, respectively. Van Veen and Paul (1981) incorporated modifications that accounted for the protection of organic matter by soil constituents, primarily clay. Under grassland, it was assumed that 50% of the organic matter was protected, whereas it was only 20% protected under cultivation.

Parton et al. (1987, 1992a, b) extended Jenkinson and Rayner (1977) and Van Veen and Paul (1981) concepts into their 'CENTURY' model for simulating the dynamics of C, N, and P in grassland and cultivated soils. Their SOM components of the model constitute an active fraction (live microbes and microbial products) with turnover periods of 1–5 years; a 'slow' fraction which is physically protected and/or in chemical structures, resistant to biological decomposition, with turnover periods of 20–40 years; and a 'passive' fraction that is chemically recalcitrant and that may also be physically protected, with turnover periods of 200–2000 years. Plant residue, above- and below-ground plant biomass, was divided into structural pools, with turnover periods of 1–5 years, and metabolic pools, with turnover periods of 0.1–1 year prior to transfer into SOM components. The above-ground organic residues also sustain microbial biomass at the soil surface.

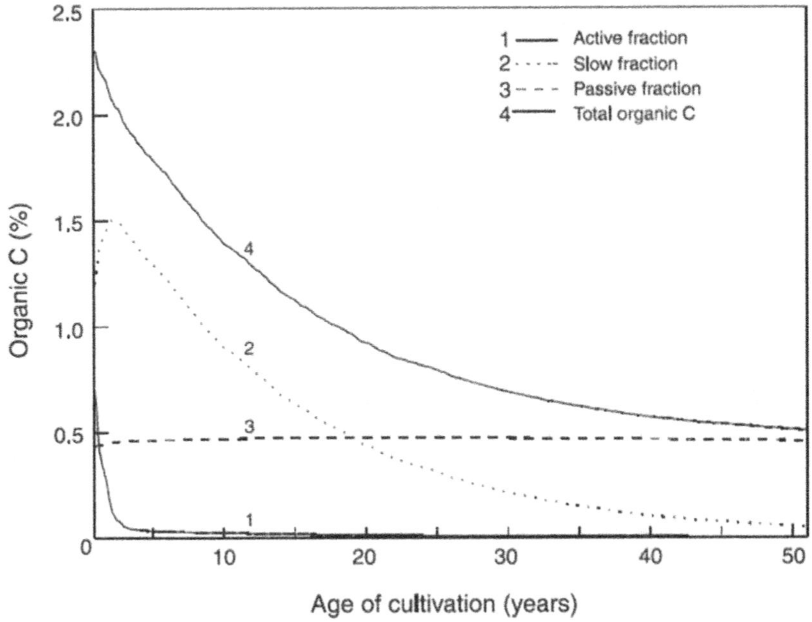

Fig. 4.4 CENTURY model simulation of soil organic C changes in a Vertisol (Langlands-Logie clay soil series) subjected to continuous cereal cropping for 50 years. (Source: Dalal and Chan 2001)

The application of the 'CENTURY' model to SOM dynamics in the Langlands-Logie soil series under continuous and cereal cropping is presented in Fig. 4.4. The active, slow, and passive components of SOM were light (<2 Mg m^{-3}) or sand-size fraction, heavy fraction 2–2.4 Mg m^{-3}, and heavy fraction >2.4 Mg m^{-3}, respectively. The magnitudes of decline in total SOC and the slow component were reasonably close to the experimentally determined values (Dalal and Mayer 1986d; R.C. Dalal, unpublished data), although the dynamics of the active component of SOM (Dalal and Mayer 1986c, 1986d) were poorly simulated.

The models of van Veen and Paul (1981), 'CREAMS' (Knisel 1980) and 'Perfect' (Littleboy et al. 1989), also incorporate the effects of soil erosion, primarily using USLE (Sect. 3.1), on SOM losses, and thus, these predict SOM level more satisfactorily in soils subjected to erosion than the 'Rothamsted' and 'CENTURY' models (Jha et al. 2021). The listing of other models relevant to SOM studies is given by Barnwell et al. (1992). Of late, various models such as APSIM, EPIC, and C-QUEST also predict carbon pools/carbon dynamics in agricultural soils in different soil types, climatic conditions, and management practices (McGill et al. 1996; McCown et al. 1996; Keating et al. 2003; Gaydon et al. 2017; Mohanty et al. 2020; Jha et al. 2021).

The SOM models were generally developed and validated in temperate regions. Parton et al. (1989) discussed at least three deficient areas, understanding which could make the applicability of these models to the SOM of tropics and subtropics more appropriate. These include (1) functional relationships of soil texture, clay mineralogy, and especially Fe and Al on SOM dynamics, (2) clay mineralogy and parent material on the formation of 'passive' SOM fractions, and (3) low pH and Al toxicity on microbial transformations. Furthermore, the modifications of the 'CENTURY' model—such as considering the role of fauna (e.g. ants, termites, and earthworms) in SOM dynamics; depth distribution of SOM <0.3 m depth; solution chemistry of SOM, especially in Ultisols and Oxisols; and, above all, aggregation dynamics of SOM turnover—are all important. Recently, a numerical model based on microbial-mediated soil aggregate formation has been developed (SAMM version 1.0), which simulated soil organic C, litter C, and aggregate C dynamics satisfactorily (Laub et al. 2024).

As mentioned earlier, there is an urgent need to experimentally verify the various 'functional' components of SOM used in these models. Most SOM models assume a 'protected' or 'recalcitrant' fraction of SOM which usually correlates with the clay (and silt) fraction, although the nature and magnitude of aggregate-size distribution in the soil would be more relevant (Elliott 1986; Dalal and Mayer 1986b; van Veen and Kuikman 1990) to SOM dynamics than the total clay (and silt) contents, especially in conjunction with the density fractions of SOM (Wang et al. 2004). Unfortunately, methods used to fractionate SOM in different aggregate-size and density fractions are poorly described, and thus, comparative evaluation of the importance of aggregate-size and density distribution of SOM on its dynamics is made extremely difficult. It is imperative, therefore, to standardise the methods of determining the fractions of SOM based on soil aggregate size and density and then to ascertain whether aggregate-size and density fractions of SOM from long-term cultivated soils are biologically meaningful in SOM turnover in subtropical and tropical soils (Dalal 1998a, b, 2021; Wang et al. 2004).

Accounting for the density fractionation, aggregate-size and particle-size fractionation led to the three main SOM pools: free particulate organic matter (fPOM), aggregate-occluded particulate organic matter (oPOM), and silt+clay size or mineral-associated organic matter (MAOM). Plant inputs in rhizosphere/litter release water-soluble dissolved organic matter (DOM), which interacts with either mineral surfaces and forms MAOM or is used by microbes and produces microbial products, which interact with mineral surfaces and form MAOM. Partially dissolved exudates can be directly involved in forming and binding aggregates or used by microbes and their products and plant fragments or hyphae acting as a nucleus to produce aggregates or oPOM. The remaining partially decomposed and fragmented plant fragments, litter, or roots and hyphae form fPOM. However, there is a continuous transformation of fPOM and oPOM into MAOM by microbial processes, including the release of simpler organic substances, and the formation of MAOM, including the release of CO_2 during the transformation process (Jones et al. 2023). Long-term stabilisation of SOC primarily occurs by two mechanisms: mineral-organic matter association or MAOM and pyrogenic organic matter or biochar (Kleber et al. 2015). This further complicates the SOC fraction protocol since pyrogenic organic matter functions differ from those of MAOM.

Robertson et al. (2019) proposed the MEMS (microbial efficiency-matrix stabilisation framework) model to utilise POM and MAOM fractions of SOM to account for SOM formation and persistence in ecosystems. Further improvement in SOM dynamics awaits the understanding of microbiome (archaea, bacteria and fungi, and their metabolic products, necromass, and enzymes) precise habitat in different pores of different sizes within aggregates in situ where the formation, stabilisation, and turnover of the aggregates and organic matter occur in soil (Schluter et al. 2022; Li et al. 2024).

Operationally, the organic matter that floats in the heavy-density liquid, either sodium iodide (NAI) solution or sodium polytungstate (SPT, $Na_6[H_2W_{12}O_{40}]$ or $3Na_2WO_4.9WO_3.H_2O$) at 1.8 Mg m^{-3} (1.8 g cm^{-3}) or <1.8 Mg m^{-3} density SOM fractions, is considered as fPOM. After separation of fPOM, occluded particulate organic matter within aggregates, oPOM and MAOM are usually dispersed either using ultrasonic energy or sodium hexametaphosphate ($Na_6[(PO_3)_6]$ solution to disperse the soil particles. In the former, ultrasonic energy, ~400–700 J mL^{-1} is applied to the soil in the SPT solution (1.8 Mg m^{-3}) contained in a temperature-controlled container. Excess oPOM is separated, and the remaining soil is sieved through a 53-μm sieve to collect <53 μm as MAOM. The >53 μm fraction remaining on the sieve is considered as the SOM in the sand-size fraction, which usually contains a small amount of SOC and could be added to the fPOM fraction. This is circumvented when after separation of fPOM, excess SPT is washed from the remaining soil and the soil is dispersed in sodium hexametaphosphate and sieved through a 53-μm sieve to collect <53 μm as MAOM. The >53 μm fraction remaining on the sieve is considered as the organic matter occluded in the sand-size aggregates or aggregate organic matter since sand fraction as such contains only a small amount of organic matter. Thus, SOM (C and N) is separated into three fractions: fPOM, oPOM, and MAOM. Obviously, dissolved organic matter (DOM) is either lost or added to the MAOM fraction, although in most mineral soils, DOM constitutes <2% of SOC (Poeplau et al. 2018; Lal 2016, 2020a,b; Lavallee et al. 2020; Cotrufo and Lavelle 2022).

4.1.5 Magnitude of SOM Loss from Cultivated Soils

The proportion of organic C lost during a given period of cultivation following land clearing can be calculated from Tables 4.1, 4.2, and 4.3 for total organic C and Table 4.5 for various fractions of SOM. The time required for half of SOM loss between initial and steady-state values, that is, the half-life is ln 2/k, so the half-life of SOM and its various fractions vary from 0.6 years to about 25 years in tropical and subtropical soils. The ^{13}C natural abundance studies corroborate these SOM half-life values (Vitorello et al. 1989; Martin et al. 1990; Dalal et al. 2011a).

Bouwman (1989) showed that SOM decomposition in the tropics was three to four times faster than in temperate regions and that both soil type (presumably soil texture) and soil environment influenced SOM decomposition. Simulated losses of organic-C in the tropical soils, without erosion losses, were 31% to 50% after

50 years and 43% to 63% after 100 years of continuous cultivation without assuming any steady-state SOM. The simulated loss of SOM when rainfall erosion was also considered was 40–80%. Soil erosion caused an extra loss of at least 7% after 100 years. As expected, fine-textured soils were less prone to SOM decomposition than soil low in clay. These simulated SOM decomposition trends are similar to those of Jenkinson and Rayner (1977) and Parton et al. (1987, 1992a, b), although SOM decomposition in many tropical and subtropical soils is much faster than the estimates of Bouwman (1989).

Russell (1984) measured a decline in organic C of 38% in a Vertisol in the subtropical and semi-arid regions after 14 years of continuous cultivation for wheat, and Standley et al. (1990) measured a 27% loss in soil organic C from a similar soil after 7 years of continuous cultivation for sorghum. Lal and Kang (1982) reported a 53% loss in organic C from a tropical Alfisol after 8 years of cultivation. Srivastava and Singh (1989) measured a 57% loss in soil organic C after 15 years of cultivation, and Martin et al. (1990) recorded similar losses, using ^{13}C natural abundance measurements, 59% after 16 years from a tropical Oxisol. Skjemstad and Dalal (1987) measured 61% of organic C loss from a Vertisol (Langlands-Logie soil series) after 35 years of continuous cultivation, whereas from a Brazilian Oxisol, organic C loss, using ^{13}C natural abundance, was estimated to be 50% after 12 years of cultivation (Cerri et al. 1985).

Organic C loss from particle size <0.05 mm was generally small (43–48%) compared with the larger particle-size fractions, <0.5 mm (89–97%) in a tropical Oxisol after 16 years (Martin et al. 1990). Skjemstad and Dalal (1990) measured almost complete loss of organic C from <2.0 Mg m^{-3} fractions, 52% from 2.0–2.4 Mg m^{-3} fraction, and none from >2.4 Mg m^{-3} fraction, with an overall loss of 61% from a Vertisol after 35 years of cultivation.

Therefore, the magnitude of not only total SOM loss but that of various fractions varies widely even in soils of the subtropical and tropical regions, and thus, the consequences of SOM loss on soil fertility depletion vary in their effects and magnitude considerably (Dalal et al. 2000; Weismeier et al. 2016; Tang et al. 2019). Wei et al. (2014) reported that the percent decrease in SOC stocks and the turnover rate constants both varied significantly according to forest type and cultivation stage. Unexpectedly, the largest decrease in SOC stocks was observed in temperate regions (52% decrease), followed by tropical regions (41% decrease) and boreal regions (31% decrease).

4.1.6 Consequences of SOM Loss from Cultivated Soils

The contributions of organic matter to soil fertility include one or more of the following aspects:

1. SOM is a major natural source of plant nutrients, e.g. N, P, and S.
2. It is a major C and energy source for microorganisms.

3. It contributes significantly to the cation exchange capacity of the soil, especially in low-activity clay.
4. SOM is an important sink of nutrients that would otherwise be lost through leaching in coarse-textured soils.
5. It provides buffering capacity to soil, thus reducing the deleterious effects of low pH and high acidity.
6. It provides polyvalent anions to complex strongly with polyvalent metal cations, e.g. Fe, Al, and the transition metal cations, thus not only providing micronutrient metal cations in plant available form but also detoxifying heavy metal cations, especially in acidic Alfisols, Oxisols, and Ultisols.
7. SOM also reduces the adsorption of P (and S) by soil colloids.
8. It promotes soil aggregation and thus improves water infiltration and water use efficiency and reduces soil erosion and to some extent wind erosion.
9. By actively reacting (or absorbing) pesticides and by providing microbial biodiversity to degrade pesticides, it reduces environmental pollution.
10. SOM provides a significant sink for atmospheric C and thus plays an important role in the 'greenhouse' effect and climate change mitigation.

Salient features of these aspects of SOM towards soil fertility degradation are outlined in Chaps. 2 and 4 of this book.

4.1.6.1 Source and Sink of Nutrients

In low-input cropping systems in the tropical and subtropical regions, SOM is the major supplier of nutrients, and it is also an important reservoir for mobile nutrients that would otherwise leach from soil, especially from Oxisols and Ultisols (Sect. 3.4). Since N, P, and S exist in SOM in relatively fixed proportions to C, e.g. C:N:P:S, 110:10:1.5:1.5 (Dalal 1977), the loss of organic C invariably leads to mineralisation and subsequent loss of these nutrients from the soil system (Mueller et al. 1985). This is discussed in detail in Sect. 4.2.

4.1.6.2 Cation Exchange Capacity (CEC)

Although humic acids and fulvic acids have very high CEC, 485–870 cmol (p+) kg^{-1} and 1400 cmol (p+) kg^{-1}, respectively (Stevenson 1982), effective CEC values for SOM are in the range of 100 to 300 cmol(p+) kg^{-1} (Greenland 1986; Logan 1990), due partly to pH dependence and blockage of negatively charged sites on SOM by interactions with Al and possibly Fe and Mn. Moreover, the SOM extraction procedures to separate humic acids and fulvic acids involve strong alkalis and acids, which would change C functional groups. Separating SOM by density and size into fPOM, oPOM, and MAOM may result in different CEC values for these fractions, which may equate to that of the whole soil. Greenland (1986) observed that CEC (determined at pH 7.0) of Ultisols and Alfisols was closely correlated with organic C content and SOM contributing, respectively, 230 and 280 cmol (p+) kg^{-1} of organic matter. Gourley (1987) found that the effective CEC of an Ultisol from Columbia decreased by 70% as organic matter decreased by 90% (from 3.55% OC at 0–0.2 m depth to 0.35% OC at 0.62–0.91 m depth), whereas in Oxisol, a decrease of

85% in OC resulted in 82% decrease in effective CEC. These studies on Oxisols are also corroborated by the studies from Brazil (Van Wambeke et al. 1983; Moraes et al. 2016. Gillman (1984) found an even more drastic effect of loss in SOM on the effective CEC of a Queensland Oxisol that had lost about 60% of SOM after 55 years of cultivation for sugarcane but lost more than 80% of its effective CEC. On the other hand, the effect of loss of SOM on the CEC of soils containing large amounts of permanent-charge clay minerals is negligible. For example, R.C. Dalal (unpublished data) found that even though the loss in SOM following long-term cultivation of Vertisols for cereal cropping exceeded 60%, CEC remained essentially unaffected in the soil.

Although the mechanisms of such effects of SOM on CEC in Vertisols and other soils are not entirely understood, it is apparent that the destruction of SOM, where it is a principal source of CEC in soil, will significantly reduce CEC levels and soil's ability to retain basic cations.

4.1.6.3 Micronutrient Availability

SOM forms relatively insoluble to soluble complexes with micronutrient cations and to some extent borate-carbohydrate complexes, which are generally considered an important labile source of these nutrient elements in soil (Dhaliwal et al. 2019, Gerke 2022). The soluble SOM-metal complexes are often a major proportion of micronutrients in soil solution, especially at high pH and in calcareous soils (Hodgson 1963) where they aid in the transport of micronutrient elements through soil solution to plant roots. Thus, micronutrient deficiencies, especially those of Fe and Zn, are found widespread in the high pH soils (Vertisols, Inceptisols) of the tropics and subtropics that have also lost considerable amounts of SOM during long periods of cultivation.

4.1.6.4 Toxic Metal Complexation

SOM also forms stable complexes with polyvalent metal cations, such as Al, Ni, Co, and Pb. This is especially important in reducing the toxic levels of these metal cations in soil resulting from strongly acidic conditions (e.g. reducing exchangeable Al from acidic Oxisols and Ultisols), industrial pollution, and industrial and mining wastes or from the application of sewage sludge containing metal wastes (Logan and Chaney 1983; Pendias and Pendias 1992). The strength of binding of the following metals to soil organic substances is usually in the order: Al > Ni > Co> Pb so that loss of SOM results in the increasing Al toxicity in acidic Oxisols and Ultisols and reduced ability to detoxify Ni, Co, and Pb, if they are added in large amounts.

4.1.6.5 Pesticide Retention

Since a large number of pesticides currently in use are aromatic and hydrophobic, they show a much stronger affinity for SOM than for water. Thus, the retention and detoxification of such pesticides in soil will be greatly decreased with the loss of SOM, by decreasing considerably the availability of organic macromolecules and polymers from SOM to deactivate the functional groups of pesticides and retain them for microbial degradation (Suhail and Fahmi 2020).

Table 4.6 Linear relationship between aggregation (y) and soil organic C (x)

Soil order	Regression constant	Regression coefficient	r value
Inceptisols[a]	−13	16.9	0.87
Alfisols[b]	−20.3	21.5	0.96
Alfisols[c]	−5.8	32.1	0.67
Vertisols[d]	72	10.5	0.80
Vertisols[e]	70	4.3	0.76

[a]Strickling (1950), [b]Tisdall and Oades (1980), [c]Chan et al. (1992), [d]Calculated from Chan et al. (1988) for aggregates > 0.05 mm diameter, [e]R.C. Dalal (unpublished data)

4.1.6.6 Microbial Biodiversity

The least is known about the changes in heterogeneity and equitability (biodiversity) of microbial population in the soil as a result of SOM loss (Sims 1990; McDaniel et al. 2014; Hartmann et al. 2015; Jayaraman et al. 2021), although it is known that microbial biomass declines with the loss of SOM [Section "Soil Microbial Biomass"] (Dalal 1998a, b; Esmaeilzadeh-Salesman et al. 2021).

4.1.6.7 Soil Aggregation and Soil Moisture Regimes

Organic matter provides the binding materials to aggregate the primary particles in the soil and thus promotes the water stability of soil aggregates (Sect. 3.6) (Jayaraman et al. 2017, 2018a, b, c, d). The decline in SOM is usually accompanied by a decrease in the number of water-stable aggregates since in most soils these are significantly correlated (Table 4.6) in spite of the fact that no standard procedure exists to measure the water-stable aggregates or extent of aggregation in soil.

The striking results from Table 4.6 are that aggregation in Alfisols (and Inceptisols) is four to seven times more responsive to the change in SOM than that in Vertisols. Dalal et al. (1991) also found that aggregation (silt and clay aggregated, <0.02 mm) declined five times faster in Alfisols than Vertisols when these soils were brought under cultivation for cereal cropping, a rate of loss in soil aggregation parallel to that of loss of SOM from these soils (Dalal and Mayer 1986b). Earlier, Marshall (1962) and Baver (1968) also observed that there was a corresponding decrease in soil aggregation as SOM declined. Also, the loss of SOM and subsequent decrease in aggregate stability occurred at a more rapid rate immediately following the cultivation of virgin soil or grasslands. This is because the transient binding organic matter (polysaccharides, polyuronides, mucilages, microbial debris) and temporary binding organic matter (roots, fungal hyphae) as fPOM or free sand-size SOM disappear at a much faster rate than the whole SOM.

Lal (2020a, b) reviewed the effects of SOM on plant available water capacity (PAWC) in some soils of the tropical region and concluded that SOM may have more beneficial effects on PAWC than the clay content. For example, Pigeon (1972) and Lal (1979) obtained the following relationships between PAWC and clay and SOM in Ultisols of Uganda (Eq. 4.2) and Alfisols of Nigeria (Eq. 4.3), respectively:

$$\text{Plant-available water } (\%) = 169.3 - 1.5\ (\%\text{clay}) + 6.09\ (\%\text{SOM}) \quad (4.2)$$

$$\text{Plant-available water } (\%) = 8.32 + 2.35\ (\%\text{OC}) \quad (4.3)$$

In low clay soils, therefore, both aggregation and PAWC are affected by SOM, and thus, its loss reduces soil aggregation and PAWC and hence leads to soil degradation, especially in Alfisols and Ultisols.

4.1.6.8 Contribution to Greenhouse Gases

Loss of SOM results in CO_2 and CH_4 emission from soil, both contributing to the greenhouse gases (Dalal and Allen 2008; Wang et al. 2011). Detwiler and Hall (1988) estimated that CO_2 emission from SOM loss was 0.11 to 0.25 × 10^{15} g C $year^{-1}$ from the tropical soils; Schlesinger (1986) and Bouwman (1990) estimated even higher values, 0.7 to 1.5 × 10^{15} g C $year^{-1}$. Sanderman et al. (2018) estimated the total SOM contribution to CO_2 emissions of 116 × 10^{15} g C since agriculture began about 12,000 years ago. The emissions of CH_4 related to agricultural activity are from rice cultivation (60–140 × 10^{12} g C $year^{-1}$), natural wetlands (40–160 × 10^{12} C $year^{-1}$), landfills (30–70 × 10^{12} g C $year^{-1}$), ruminants and termites (65–100 × 10^{12} g C $year^{-1}$), and biomass burning (50–100 × 10^{12} g C $year^{-1}$) (Bouwman and Sombroek 1990; Dalal et al. 2008). Emissions of nitrous oxide (N_2O) from agricultural soils have been steadily increasing as the N fertiliser use increases to compensate for increasing fertility depletion, which now accounts for 6% of global warming (Dalal et al. 2003; Wang et al. 2016).

4.2 Nutrient Depletion

Soil fertility degradation primarily occurs through nutrient depletion when the amounts of available nutrient elements become inadequate for plant biomass production (Dalal and Mayer 1986a, b, 1987a, b). The available nutrients when depleted are replenished from the total soil reserves through the mineralisation of organic matter (N, P, and S), dissolution of mineral precipitates (P, K, Ca, Mg, S, B. Fe, Mn, Cu, Zn), and desorption of strongly adsorbed chemical species (P, K, and trace elements). Thus, soil fertility depletion results in decreased amounts of not only the plant available nutrients but also the total nutrient elements in the soil (Dalal and Mayer 1986a, b, 1987a, b).

Loss of SOM on cultivation of virgin or pasture lands is also accompanied by the loss of its constituent nutrients, especially those that are not readily retained by the soil, for example, N and S (Dalal et al. 2000). Besides the loss of nutrients associated with loss of SOM, it can occur by one or more processes of soil fertility depletion, such as nutrient removal in produce, and plant residue, soil erosion, burning, reduction (denitrification), and leaching (Sects. 3.1, 3.2, 3.3 and 3.4). Cultivation effects on N, P, S, K, exchangeable Ca and Mg, and diethylenetriaminepentaacetic acid (DTPA) extractable Cu, Zn, and Mn in soil are briefly considered.

4.2.1 Nutrient Eelement Loss

4.2.1.1 Nitrogen

The most common nutrient element limiting plant biomass production and hence the major cause of soil fertility degradation in many soils is soil N depletion (Dalal 1992; Lal and Stewart 2018). Management of N is, therefore, crucial for sustainable crop production and restoration and maintenance of soil fertility (Dalal and Mayer 1987a, b; Schlesinger 2009; Dalal et al. 2011a; Robertson and Groffman 2024). Moreover, attaining food and nutritional security is possible not only by the adoption of sustainable intensification but also due to the application of fertiliser nutrients particularly N-based fertiliser, which has increased seven-fold in N fertilisation and double the N load in the soil at the beginning of the twentieth century (Green et al. 2004).

Soil N Contents of Natural Ecosystems

The amount of soil N content in a natural ecosystem is related to climate (biotemperature, annual precipitation, and potential evapotranspiration/precipitation ratio) through biotic processes associated with the productivity of vegetation and decomposition of organic matter (Schlesinger 2009). Other factors, particularly N in rainfall and atmospheric dry deposition, N_2 fixation and losses of inorganic N due to leaching and/or denitrification, and N removal in products and erosion, contribute to the variability of soil N contents (Post et al. 1985; Cassman et al. 2002; Lal and Stewart 2018). The estimated total soil N is 95×10^{15} g; almost 40% of this N is stored in tropical and subtropical soils (Table 4.7). Unfortunately, actual soil N storage is lower than this value because of the loss of N when soil is cultivated. The global size of such losses, however, is presently unknown (Post et al. 1985). Fowler et al. (2013) reported that global nitrogen fixation (annually) contributes 413 Tg (1 Tg$=10^{12}$ g) of reactive nitrogen (N_r) to terrestrial and marine ecosystems, of which anthropogenic activities are responsible for half, 210 Tg N. The land contributes 240 Tg N year^{-1} about 58%, and the majority of the transformations of anthropogenic N occur within soils and vegetation where reduced N contributes most of the input through the use of fertilizer nitrogen in agriculture. Smil (1999) reported that anthropogenic inputs (consisting of N in seeds and irrigation water) can provide ca 85% of 170 (151–186) Tg N reaching the world's cropland annually. From which about half of this input, 85 Tg N year^{-1} has been taken up by harvested crops (grain and residue). However, uncertainties exist in the quantification of N losses from crop fields. Similarly, the losses to the atmosphere such as denitrification and volatilization amount to 26–60 Tg N year^{-1}, and water receives about 32–45 Tg N year^{-1} through leaching and erosion.

4.2.1.2 Loss of Total N from Cultivated Soils

Since the loss of N from soil may be due to a number of factors, such as erosion and runoff, leaching, denitrification, burning, and the removal of plant residues and economic produce such as grains, the estimates of soil total N losses vary (Vitosh et al. 1995a, b; Cameron and Moir 2013). Generally, rates of soil total N losses from

Table 4.7 Soil nitrogen and C/N ratio of tropical and subtropical natural ecosystems

Ecosystem	Total soil N ($\times 10^{12}$ g)	Soil profile N (g m^{-3})	C/N ratio
Tropical thorn woodland[a]	622.1	264.6	9.2
Tropical very dry forest[a]	2814.0	597.2	13.7
Tropical dry forest[a]	8874.9	885.9	13.3
Tropical moist forest[a]	6909.0	802.9	14.9
Tropical wet forest[a]	247.7	655.2	30.2
Subtropical desert bush[a]	3503.8	2282.6	12.7
Subtropical thorn woodland[a]	626.2	379.3	14.3
Subtropical dry forest[a]	3746.4	1070.4	10.2
Subtropical moist forest[a]	7888.4	987.9	10.3
Subtropical wet forest[a]	1423.8	2853.4	3.3
Subtropical semi-arid region[b]		685.0	12.1
Grassland[b]		770.0	11.3
Savannah (Eucalypt-grassland)[b]		674.0	13.0
Acacia forest[b]		869.0	10.9
Casuarina forest[b]		710.0	10.6
Eucalypt woodland (Vertisol)[b]		540.0	11.8
Eucalypt woodland (Alfisol)[b]		546.0	15.2
Tropics and subtropics[b]	36656.3		
World	95433.7		

[a]Post et al. (1985). [b]Estimated from Dalal and Mayer (1986e). [c]Excluding tropical and subtropical deserts (559.5×10^{12} g N) and wetlands (1000×10^{12} g N)

cultivated soils can be inferred from the organic C loss (Sect. 4.1). However, it is only after long periods of cultivation that losses of soil N are proportional to the losses of organic C and that factors of nutrient depletion do not always affect equally both organic C and N (Wang and Dalal 2015; Dalal et al. 2011c; Robertson and Groffman 2024).

Overall Rate of Soil N Loss

Nitrogen is one of the most dynamic, mobile, and enigmatic nutrients in the soil (Allison 1955, 1966; Rose and Dalal 1988; Wang and Dalal 2015). Therefore, the estimates of the rate of soil total N losses from cultivated Vertisols vary widely. Martin and Cox (1956) measured 0.8% and Hallsworth et al. (1954) 2% annual decrease in soil N from the top 0.15 m layer, whereas Waring and Teakle (1960) estimated that 3% of total N was removed by wheat crops annually. Russell (1981) measured N losses of 5% annually from an initially highly fertile Vertisol cropped to sorghum continuously for 10 years. Dalal and Mayer (1986e), using Eqs. 4.1a, 4.1b, 4.1c and 4.1d, found that rates of soil total N loss increased with the decrease in the clay contents of the Vertisols studied; the rates of loss increased from 0.061 $year^{-1}$ in Waco clay (74% clay), 0.115 $year^{-1}$ in Langlands-Logie clay (49% clay), to 0.275 $year^{-1}$ in Cecilvale clay (40% clay). Thus, the clay fraction provides protection to organic matter or makes it less accessible to degradative enzymes and

Table 4.8 Annual rate of loss of N from the profiles of six soil series and estimated annual N removal by cereal crops (mean values in 1982–1984)[a]

Soil taxonomy	Soil series	Rate of loss (kg N ha^{-1} $year^{-1}$)	Rate of crop removal (kg N ha^{-1} $year^{-1}$)	Fertiliser applied (kg N ha^{-1} $year^{-1}$)
Rhodic Paleustalfs	Riverview	32.0 ± 11.0	35 ± 17	0
Typic Pellusterts	Waco	31.3 ± 3.0	59 ± 21	33
Typic Chromusterts	Cecilvale	34.5 ± 5.5	45 ± 16	18
Typic Chromusterts	Thallon	35.8 ± 8.6	42 ± 20	0
Typic Chromusterts	Billa Billa	50.8 ± 7.6	66 ± 25	0
Typic Chromusterts	Langlands-Logie	67.1 ± 6.9	51 ± 13	8

Dalal and Mayer (1986e)

microorganisms (Rose and Dalal 1988). Liu et al. (2010) reported that 55% of the global applied N was taken up by crops. The remainder N was lost, namely, leaching (16%), soil erosion (15%), and gaseous emission (14%). These depletions of N lead to eutrophication in water bodies (Stoate et al. 2009) and nitrous oxide emissions to the atmosphere (Dalal et al. 2003; Wang and Dalal 2015).

The mean annual rates of N loss from the profiles of five Vertisols and one Alfisol subjected to 20–70 years of cereal cropping were 31.3–67.1 kg N ha^{-1} $year^{-1}$ (Table 4.8), which were similar to that accounted for in crop removal. Some of the N loss was compensated for by the application of fertiliser N, which is usually 50% utilised by the crop (McDonald 1989) and removed in crop produce (Dalal et al. 2011a). Similar studies done elsewhere on soils subjected to long periods of cultivation have shown that in most cases, soil N losses were primarily accounted for in the N removal by crops (Haas et al. 1957; Williams and Lipsett 1961; Clarke and Russell 1977), especially where no organic manures or fertilisers have been added. However, in soils of initially high fertility, N losses by other factors, such as by deep leaching (Sect. 3.4) and possibly by denitrification, especially from applied fertiliser N (Moraghan et al. 1984), do occur and eventually result in soil fertility degradation.

Rate of N Loss from Various Soil Fractions

Similarly to soil organic C (Dalal and Mayer 1986c), the light fraction (<2 Mg m^{-3}) N is lost at a much faster rate than the heavy fraction (>2 Mg m^{-3}) N in Vertisols (Dalal and Mayer 1987a) and Alfisols (Ford and Greenland (1968), when these soils are cultivated. The inverse relationship between the rate of loss of N from the light fraction and the clay content (Dalal and Mayer 1987a) demonstrates that the light fraction N is probably not associated or only partly associated with the clay fraction N (mineral-associated organic matter). Therefore, unlike the heavy fraction N, which is closely associated with the clay fraction N, the light fraction N in most soils is poorly protected by clay, and hence, it is a relatively labile fraction N in the soil.

Since the sand-size fraction N is closely correlated with the light fraction N (fPOM N) (Dalal and Mayer 1987a; Christiansen 1992), it also declines at a faster

Table 4.9 Annual percent decrease in total N from sand-, silt- and clay-size particles from cultivated Vertisols and Alfisols[a]

Soil order (soil series)	Clay (%)	Mean annual decrease (% $year^{-1}$)		
		Sand-size N (>0.02 mm)	Silt-size N (0.02–0.002 mm)	Clay-size N (<0.002 mm)
Vertisols:				
Pellusterts (Waco)	74	2.7	0.2	1.2
Chromusterts (Langlands-Logie)	49	4.9	2.6	0.6
Alfisols:	18	10.5	5.9	0.1
Peleustalfs (Riverview)				

[a]Calculated from Dalal and Mayer (1987a)

rate than either the silt-size N or the clay-size N (MAOM N) upon the cultivation of virgin and pasture soils (Table 4.9). Generally, the rate of loss of N decreases as the particle size decreases. Thus, in coarse-textured Alfisols, most of the decreases of N upon cultivation occur from the sand-size and silt-size fractions and almost negligible decreases from the clay-size N. However, in fine-textured Vertisols, clay-size N also declines upon cultivation (Table 4.9) since it is a major source of mineralisable N. It is not known whether similar trends exist for changes in various fractions of N in Ultisols and Oxisols when they are brought under cultivation, especially in tropical and subtropical regions.

Since the decreases in the 'labile' N fraction could be proportionately greater than that of soil total N, the N-supplying capacity of soil could decrease much faster than that shown by total N. Dalal and Mayer (1987b) found that N contained in microbial biomass in a Vertisol declined upon cultivation at a much higher rate (0.338 $year^{-1}$), with a turnover period of <3 years, than the rate of total N loss, a result similar to that observed by Adams and Laughlin (1981), Jenkinson and Ladd (1981), and McGill et al. (1986) for other soils and in different climates.

Modelling of Soil N Turnover

Of late, the use of advanced molecular tools/techniques (Gabriel and Quemada 2011) and stable isotopes (Baggs 2008) helps in deciphering the vital regulators of N turnover for selected processes such as nitrification (Leininger et al. 2006) or N fixation (Chowdhury et al. 2009) as well as processes like anammox, i.e. anaerobic ammonia oxidation (Op den Camp et al. 2006).

The soil N turnover generally parallels that of SOM turnover, which has been described in Sect. 4.1.4, and therefore, only a few features of soil N turnover will be briefly mentioned. In a natural ecosystem, the turnover period for the global soil N pool is estimated to be 3000 years (Post et al. 1985), although the turnover periods of various N fractions within the total N pool vary from weeks to a few thousand years (Jenkinson et al. 1991, Parton et al. 1990). Except for the N loss by en masse soil removal, most of the N losses from soil occur from N fractions with turnover periods, varying from weeks to a few years, especially when these are not synchronised with the N demand by plant biomass. Current models inadequately describe the turnover rates of these 'labile' N fractions in soil (Chilcott et al. 2007).

Consequences of Soil N Loss

Since continuous cultivation and cereal cropping of soil initially high in fertility lead to the decline in organic matter and its associated nutrient constituents, such as organic N (Dalal and Mayer 1986a), a decline in cereal grain protein occurs (Martin et al. 1988; Dalal et al. 1991), without supplementary N, as the period of cultivation increases (Fig. 4.6 (24)). Furthermore, cereal grain yield also declines with cultivation (Dalal 1990) because of soil nutrient depletion (Fig. 4.6 (24)), especially plant available N. For example, wheat cereal grain yield declined from more than 4 Mg ha^{-1} up to 5 years to less than 2 Mg ha^{-1} after 30 years of cultivation of virgin Waco (Typic Pellustert) and Billa Billa (Typic Chromustert) soils. Grain yields were closely correlated with mineralisable N and nitrate N at planting (Dalal and Mayer 1990). From their survey studies, Martin et al. (1988) also found that wheat grain yields and protein contents were significantly correlated with soil nitrate N at sowing, and they estimated that the optimum wheat grain yields in the field were close to 5 Mg ha^{-1} in the subtropical environment, although for summer crops, in-crop N mineralisation may account for a significant contribution to total crop N uptake.

Crop productivity decline in Alfisols and Ultisols is more rapid than in Vertisols (Lal 1990; Obi 1989; Dalal and Mayer 1986a), mainly because of low initial nutrient concentrations and rapid leaching losses in cultivated soils. Also, since plant nutrients are relatively more concentrated near the soil surface in Alfisols and Ultisols than in Vertisols, even moderate soil erosion from the former can reduce crop yields dramatically (Rose and Dalal 1988a, 1988b).

4.2.1.3 Phosphorus

The total P content of the soil is primarily determined by the parent material and later modified by the soil-forming processes (Smeck 1985). Vertisols vary in their total P content. Vertisols formed on some basalts have very high P content, whereas on other basalts and with many of the sedimentary rocks, total P content is low (Dalal 1979; Finck and Venkatachalam 1982; Hubble 1984; Probert et al. 1987). Similarly, Alfisols differ considerably in their P content. However, inorganic P forms in these two groups differ. Whereas, in Vertisols, inorganic P is predominantly associated with Ca, in Alfisols, it is mainly associated with Al and Fe. Organic P as a proportion of total P varies from 20% to 80% (Dalal 1977, 1979). Organic P in five Vertisols from southern Queensland varied from 21% in Waco clay to 50% in Cecilvale clay; in Alfisols, it was 27% (Dalal and Mayer 1986a). Organic P content in virgin Vertisols and Alfisols was closely associated with their organic C content. Sodium bicarbonate-extractable P values are generally higher in Vertisols (14–51 mg P kg^{-1} soil, 4–10% of total P) than in Alfisols (8 mg P kg^{-1} soil, 2–3% of total P). Phosphorus contained in microbial biomass, which contributes to plant available P upon mineralisation, comprises 2% of total P in Vertisols (Dalal 1979; Saffigna et al. 1989) and about 5% of organic P in Vertisols and other soils (Dalal and Mayer

Table 4.10 Rates of loss of organic P, organic S, and DTPA extractable Mn, Cu and Zn from cultivated Vertisols (Waco soil series)[a]

Constituents	Initial concentration (mg kg^{-1})	Steady-state concentration (mg kg^{-1})	Rate of loss (k $year^{-1}$)	Turnover period (year)
Organic P	147	71	0.033	30.3
Organic S	216	127	0.064	15.6
DTPA extractable Cu	1.52	1.22	0.072	13.9
DTPA extractable Zn	0.88	0.50	0.104	9.6
DTPA extractable Mn	27.2	13.3	0.155	6.5

[a]R.C. Dalal, unpublished data

1987b). Blake et al. (2000) reported from a 30-year long-term study that soil type had a significant influence on the dynamics of P (Chatterjee et al. 2022). The study findings suggested that variations in the soil properties (physico-chemical, biological), other major nutrient availability, the forms of P applied, and climate all significantly affect the P balance and effectiveness of P fertilization in soil.

Rate of Loss of P from Cultivated Soils

Phosphorous plays a significant role in achieving food grain production; however, depletion of P or loss of P not only impacts food production but also the environment. Globally, in arable soils, about 5.9 kg ha^{-1} P loss occurs due to erosion by water (Alewell et al. 2020). Different regions experience different magnitudes of P loss, e.g. Europe (<1.2 kg ha^{-1}), Africa (9.7 kg ha^{-1}), and South America (6.1 kg ha^{-1}).

Since organic P is a constituent of organic matter, the decline in the latter with cultivation in Vertisols, Alfisols, and other soils also leads to a decline in the soil organic P content. For example, in Waco clay, organic P declined at the rate of 0.033 $year^{-1}$ (Table 4.10). This is about half the rate of loss of organic C, 0.065 $year^{-1}$ (Dalal and Mayer 1986b), from this soil, although in Cecilvale clay, the rates of losses of organic P, C, and N vary much less (0.18–0.28 $year^{-1}$), thus indicating that the forms of organic P may differ in these soils. In another study, McDowell and Monaghan (2002) reported from a lysimeter experiment that the major form of P leached is dissolved organic P (DOP), which represented 67 and 80% of total dissolved P (TDP) losses in the undisturbed and cultivated soils, respectively.

The plant available P pool in Vertisols is satisfactorily measured by sodium bicarbonate extractable P or some other suitable technique such as an anion-exchange membrane. It is therefore likely that the available P pool in the soil would be depleted by crop removal and other P-depleting factors, such as leaching, erosion, and runoff (Djindjic et al. 2004; Verbree et al. 2010). However, there is a scarcity of information available on this aspect from the semi-arid, subtropical regions. Djindjic et al. (2004) reported that P loss from arable soils contributes to surface water pollution (i.e. eutrophication, P enrichment in the aquatic ecosystem); surface runoff, leaching, and sometimes preferential flow are the major factors contributing to the loss of P in soil or water transport pathway. Kronvang et al. (2007) reported that mean annual P loss from agricultural land in the micro-catchments and macro-catchment varied from 0.1 to 4.7 kg P ha^{-1} and 0.1–6.0 kg P ha^{-1}, respectively, but the value was less for the ten larger river basins (i.e. 0.09–2.0 kg P ha^{-1}). They used the MONERIS model and simulated the main diffuse pathways for total P loads under 17 macro-catchments and found that, on average, soil erosion and surface runoff have contributed 53% (4.1–81%); groundwater, 14% (0.2–41.7%); and tile drainage water, 3% (0–14.0%).

Chan et al. (1988) measured substantially lower (39–71%) available P, using Bray No.1 extractant, in a number of Vertisols (Western New South Wales) cultivated for 8–50 years than those under permanent pasture. They ascribed the decrease in available P due to the increase in soil pH upon cultivation. Standley et al. (1990) also found that the available P (bicarbonate extractable) in an Entic Chromustert from Central Queensland decreased from 31 mg P kg^{-1} soil initially to <20 mg P kg^{-1} soil after 7 years of cropping for sorghum, but they found no significant change in soil pH. R.C. Dalal (unpublished data) measured changes in total P, organic P, and bicarbonate extractable P in a number of Vertisols and Alfisols subjected to continuous cultivation and cereal cropping for 20–70 years. Mean annual rates of decline in total P, organic P, and bicarbonate extractable P during 45 years of cereal cropping were 8.4 mg P kg^{-1}, 3.0 mg P kg^{-1}, and 1.2 mg P kg^{-1} soil, respectively. Assuming the grain P concentration of 3.4 kg Mg^{-1} (Table 3.3), all of the soil total P removed in grain is equivalent to the grain yield of 2.5 Mg ha^{-1} $year^{-1}$ or 1.95 Mg ha^{-1} $year^{-1}$ if the crop residues are not retained. Soil organic P would have contributed 36% of the total P removal by the crop, a rate of decline similar to that of the total P. The bicarbonate extractable P, which is derived from both organic P and inorganic P in soil (Dalal and Mayer 1986a), below 10 mg P kg^{-1} indicates likely crop response to fertiliser P after about 40 years of cultivation in this soil. In general, reduction in P fertility due to the loss of 'labile' P has been closely associated with SOM losses (Tiessen et al. 1983; Dalal and Chan 2001; Maranguit et al. 2017).

Generally, P is strongly bound by soil particles. Soil erosion is the dominant process where a large portion of P is lost (Basden et al. 2023), whereas dissolved P can be easily washed away by surface runoff from cultivated or un-tilled soils. Daryanto et al. (2017) conducted a meta-analysis of P loss from conventional (CT) and no-tillage (NT) practices. Particulate P loss was significantly lower with NT adoption which translates to a 45% and 55% reduction in concentration and load, respectively, in comparison with the CT practice. However, NT increased the loss of dissolved P, possibly due to the interacting effects of different physical factors such as climate, rainfall variability, transport pathway, slope gradient, and management variables such as NT duration and crop species.

It has been suggested that inorganic P in soil solution is replenished by mineralisation of microbially derived diester-P, including phospholipids (Condron et al. 1990) and nucleic acids (Dalal 1977), which was confirmed by using ^{31}P

nuclear magnetic resonance (^{31}P NMR) techniques (Zhang et al. 2021). However, information is scanty on the rates and magnitude of various forms of organic P as well as inorganic P loss from cultivated soils of many tropical and subtropical regions. Furthermore, the role of SOM in contributing towards and maintaining plant available P in the soils of these regions has not been well documented. Of late, the recent study by Alewell et al. (2020) indicated that agricultural soils will have a depletion of soil P (4–19 kg ha^{-1} year^{-1}), assuming that there is a shortage of mineral fertiliser supply. However, about 50% of total P is lost through soil erosion, thus emphasizing the importance of soil protection against erosion.

Modelling of Soil P

A complementary approach to P management in the soil is the modelling of P dynamics under cultivation and cropping. In one of the currently used models, the 'CENTURY' model (Parton et al. 1992a, 1992b), five pools of inorganic P are assumed: labile P, sorbed P, parent P, secondary P, and occluded P. Physical and chemical weathering transform primary P (mainly apatite P) to labile P, which is then either transformed into secondary P or taken up by plant and soil organisms. As more P is removed by uptake by plant and soil organisms, larger amounts become immobilised as organic P in SOM. However, the P dynamics of the CENTURY model has not been validated for many subtropical and tropical soils. In relatively high P fixing soils, there is a large proportion of soil P which can be fixed into an insoluble form as occluded P, as in many Oxisols. The weathering rate for parent P is assumed to be a function of soil texture (higher for fine-textured soils), and the flow from secondary P to labile P is a function of the soil pH (higher for high pH soils) and soil texture (lower for fine-textured soils). The organic P of the P submodel is similar to that of C (Figure 4.4) and N, with the C:P ratios of organic fractions being fixed for the structural P pool (500) and varying as a function of the labile P pool for the active fraction (30–80), slow fraction (90–200), and passive fraction (20–200) of SOM. The C:P ratios of newly formed surface microbial biomass are assumed to be a function of the P content of the plant material (structural, 10–50; metabolic, 500) that is being decomposed, that is, higher in Oxisols than in Vertisols.

The losses of P from the soil system occur due to the leaching of labile P (high in coarse-textured soils), soil erosion, crop removal, grazing, and burning. Since there is a lack of an adequate database to initialise the P state variables for many tropical and subtropical soils, the CENTURY model has not been successful in simulating P dynamics in these soils (Parton et al. 1989; Reid et al. 2018; Das et al. 2019). There is, therefore, an urgent need to understand the soil P turnover in tropical and subtropical regions, especially for sustainable crop production on Oxisols as well as on Alfisols and Ultisols.

DSSAT model is also able to predict the P dynamic in soil. This model proposes three fractions, namely, labile, active, and stable soil inorganic P (Pi) (Dzoti et al. 2010). Reid et al. (2018) suggested that reducing the P losses from agricultural land requires a thorough understanding of the forms (i.e. particulate P, dissolved P, and chemical speciation), sources (i.e. soil erosion, desorption of soil P, dissolved P from fertilizer, or manure application), and transport pathways (surface runoff or sub-surface runoff through tile drains). They proposed a risk assessment tool, namely, the Canadian National Indicator of Risk of Water Contamination by Phosphorus (IROWC-P). Many models are available, yet the prediction P using these models is challenging. Application of this model or any other model for P dynamics in the fPOM, oPOM, and MAOM has not been undertaken for the subtropical and tropical soils.

Unlike C and N models, modelling of soil P always poses a challenging task due to the following reasons, as suggested by Das et al. (2019). Three major constraints are (i) *a scarcity of field datasets/information to model strategies/approaches aimed at improving P use efficiency, (ii) a limited ability to predict P cycling and availability under the climate change effects (local effects), and (iii) inability to match determined soil P fractions versus conceptual and modellable pools in soils with varying mineral properties.* Similar to SOM, P dynamics in fPOM, oPOM, and MAOM is required for different land uses and soil types in these regions.

Effects on Crop Yields

Phosphorus deficiency occurs worldwide, for example, Sanchez and Salinas (1981) reported that about 82% of tropical soils are deficient. However, MacDonald et al. (2011) estimated that 29% of croplands globally are deficient in P. Of late, Alewell et al. (2020) reported that agricultural soils will be reduced in P by 4–19 kg P ha^{-1} $year^{-1}$, indicating that the average loss of P owing to soil erosion is 50% of total P. Depletion of available P from continuous cultivation and cropping limits crop yields in many instances. In fact, Grundon et al. (1985) found from the pot experiments that P deficiency was widespread in southern Queensland soils, although it was found less common in field experiments, thus indicating the potential for P deficiency in this region. On Alfisols and Ultisols, however, the P deficiency is the constraint to crop production (Williams and Probert 1984; Sanchez 1987). Smith et al. (2022) reported an 80% decrease in aboveground biomass due to water stress and P deficiency either individually or the combined effect. Moreover, decreased photosynthesis and stomatal conductance occurred during plant development due to water deficit and P deficiency. In another study conducted by Yin et al. (2015), grain yield was reduced by 38.3% due to P deficiency; besides a significantly reduced initial density, limited plant growth, and exacerbated density reduction, a balanced application of NPK is required to overcome yield loss. Moreover, management practices including NT, fertiliser placement (broadcast, point placement, sub-surface injection), and cover crops play a significant role in reducing P loss in soil (Beegle and Durst 2002; Panagos et al. 2022; Carver et al. 2022). McDowell et al. (2024) emphasised the threshold level for getting optimal yield. They studied the plant available soil Olsen P concentrations versus thresholds for optimal productivity across 28 of the world's most widely grown and valuable crops and grasslands. They found that > 73% of farmlands were below optimal production thresholds; thus, judicious use of P fertiliser is required for optimal yield as well as reducing P loss in soil.

4.2.1.4 Soil Sulphur

Sulphur is present mostly (>90%) as organic S in the top 0.1 m soil layer (Bettany et al. 1979; Singh et al. 2004). However, in the sub-soil layers of Vertisols in the semi-arid regions, inorganic S, primarily as Ca (Mg) SO_4 precipitate or crystals, may exceed more than 50% of the total S. The contents of total S in the top 0.1 m soil layer vary not only among Vertisols (215–373 mg S kg^{-1} soil) but also between Vertisols and Alfisols, the latter containing less total S (119 mg S kg^{-1} soil) than the former (Dalal and Mayer 1986a). Of late, decreases in S content/deficiencies of S in agriculture are more prevalent due to S removal in the produce and unbalanced NPK fertilisation. However, very limited information is available on the long-term cultivation, availability, and S pools in soil (Dick et al. 2008; Kopittke et al. 2016). It has been reported that long-term cultivation (up to 70 years) led to a 35–51% decrease in total S due to the mineralisation of organic S and the removal of produce. Thus, these soils need to be managed sustainably with proper S nutrient replenishment (Kopittke et al. 2016).

Rate of Loss of Soil S from Cultivated Soils

The mineralisation rate of organic S corresponds to that of organic C and organic N in the soil. With the decline in organic C and N in soil under continuous cultivation and cereal cropping, organic S also declines (Dick et al. 2008; Kopittke et al. 2016). For example, in a Waco clay (Typic Pellustert) cultivated for 0–70 years, the rates of loss of organic S, organic C, and N were 0.064 $year^{-1}$, 0.069 $year^{-1}$, and 0.061 $year^{-1}$, respectively (Dalal and Mayer 1986b, c; R.C. Dalal, unpublished data, Table 4.10). Organic S declined from 216 mg S kg^{-1} soil initially to 128 mg S kg^{-1} soil after 70 years of cultivation. Again, in the Langlands-Logie clay (Typic Chromustert), the rate of loss of organic S (0.105 $year^{-1}$) was similar to that of total N (0.09 $year^{-1}$) but slightly higher than the organic C loss (0.06 $year^{-1}$). Respective mean annual rates of loss of organic S from Waco and Langlands-Logie clay are calculated to be 1.25 kg S ha^{-1} for up to 70 years and 4.7 g S ha^{-1} for up to 45 years of cultivation, respectively. Williams and Lipsett (1961) found that approximately 30% of organic S from Alfisols was lost when soils were cultivated for wheat for 50–60 years. Standley et al. (1990) measured a decline in total sulphur of 37 kg S ha^{-1} during 5 years of cultivation of a Vertisol.

An annual decline in organic S of 4.7 kg S ha^{-1} in Langlands-Logie clay can be accounted for by the removal of wheat grain of 2.6 Mg ha^{-1} (1.8 kg S Mg^{-1}, Table 3.3), provided that losses of organic S below the top 0.1 m layer are negligible. The grain yield value calculated for S loss is similar to that of total P loss from this soil.

Sulphate S, usually phosphate extractable S, is a measure of available S in the soil. The decline in sulphate S below 5 mg S kg^{-1} soil is associated with the decrease in grain yield, especially when sub-soil S concentration is low, such as the deeply leached shallow soils (Loader and White 1979), acidic Alfisols, and Ultisols. In Vertisols, such as in Waco clays, the decline in sulphate S upon continuous cultivation and cereal cropping was inconsistent and small, although organic S declined by 40% in 70 years, indicating thereby the rapid replenishment of sulphate S after its crop removal (Kopittke et al. 2016). There was a general trend in decline in sulphate S of 0.3–0.5 mg S kg^{-1} soil $year^{-1}$ in a Langlands-Logie clay (Vertisol) and Riverview sandy loam (Alfisol) subjected to 20–45 years of cereal cropping. In the latter soil, low sulphate S (<10 mg S kg^{-1} soil) after 15 years of cultivation may be limiting yields of cereals.

The 'CENTURY' model also simulates S dynamics in cultivated soils based on a modelling structure similar to that of P (Parton 1994; Parton and Rasmussen 1994). The only major difference is that the S model does not include either an occluded or a sorbed S pool. Unfortunately, the performance of the S component of the CENTURY model has not been examined extensively (Parton 1994; Parton and Rasmussen 1994), and thus, its applicability and limitations for S dynamics in tropical and subtropical soils are less known. Similar to SOM, N, and P, S dynamics in fPOM, oPOM, and MAOM is required for different land uses and soil types in these regions.

4.2.1.5 Soil Potassium

Potassium is one of the key major nutrients for increasing food production. Due to intensive agriculture, soils are showing declining status in K content. Brownlie and co-workers (2024) reported that about 20% of agricultural soils are showing acute K deficiency worldwide. Similarly, the South Asian region recorded higher K deficiency (44%), followed by Latin America (39%), sub-Saharan Africa (30%), and East Asia (20%). In another study, Wang et al. (2024) reported that cereal-based cropping systems depleted more K in soil (~62%); therefore, they suggested that replenishing K in deficient soils could potentially enhance crop yield by ~50%. Moreover, N and P applications might be masking the importance of K, but it is unequivocally important to sustain global food production and security (Brownlie et al. 2024; Wang et al. 2024).

Soil K exists in four forms, namely, solution K, readily available K, less readily available K, and non-available K (Fig. 4.5). Total K and exchangeable K in Vertisols vary (Sekhon et al. 1992; Srinivasarao et al. 2007). For example, in five Vertisols and one Alfisol soil series, Dalal and Mayer (1986a) found that total K varied from 0.30% to 0.93%, whereas exchangeable K varied from 0.7 c(p+) kg^{-1} to 2 c(p+) kg^{-1}. Only in one Vertisol, Langlands-Logie clay, significant decreases in total K (23%) as well as exchangeable K (46%) were found when these soils were cultivated for 45 years (Dalal and Mayer 1986a). The rates of decline of total K and exchangeable K, 0.108 $year^{-1}$ and 0.121 $year^{-1}$, respectively, were essentially similar to total N and total S in this soil.

Rate of Loss of Soil Potassium from Cultivated Soils

Total K declined at the rate of 42 kg K ha^{-1} in the Langlands-Logie clay during 45 years of cultivation. Assuming an annual mean wheat grain yield of 2.5 Mg ha^{-1} and K concentration of 4.5 kg Mg^{-1} (Table 3.3), the removal of K in grain amounts to only 11 kg K ha^{-1}. However, the removal of crop residue, assuming a harvest index of 0.4 and a K concentration of 9.1 kg Mg^{-1} (Table 3.4), is estimated to be 34 kg K ha^{-1}, a total removal in grain and crop residue of 45 kg K ha^{-1}. Therefore, it

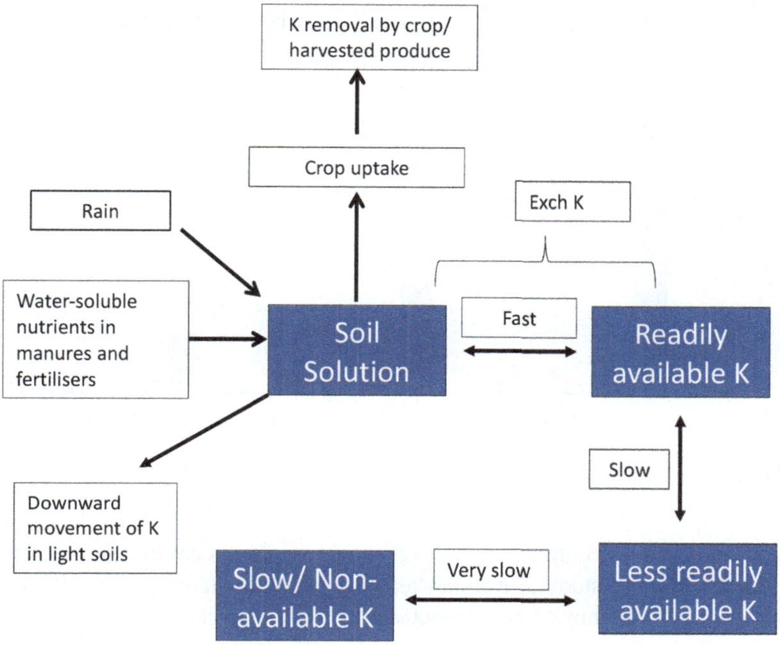

Fig. 4.5 Forms of potassium in soil. (Source: authors developed)

emphasises the retention of crop residue to minimise the loss of K from soil. Standley et al. (1990) also measured similar rates of loss in total K (50 kg K ha^{-1}) from an Entic Chromustert during 5 years of cereal cropping for sorghum in Central Queensland; however, no significant decline in exchangeable K was observed. The K availability is sufficient in most soils for arable cropping except sugarcane and other high-K-demanding crops, however, it is one of the most neglected nutrient elements (Sardans and Peñuelas 2015). Considering that most alluvial soils are rich in K, which is frequently not replenished to compensate for the depletion or uptake by crops, as a result, K deficiency occurs in many soils (Singh et al. 2019). In fact, in the LTFE, there was a response for the application of K in wheat and soybean crops in Vertisols of central India. Singh and Wanjari (2012) reported that the absence of K in the fertilizer schedule led to a decline in K status at the rate of 2.1 to 9.7 kg ha^{-1}, which was further depleted with the application of P. On the other hand, the application of K helped in arresting the decline in available K and sometimes simultaneously increased the available K form. The results highlighted that there is a need to alter or raise the K limits for rating the Vertisols as high status and accordingly K recommendation be revised in Vertisols. The study conducted by Pettigrew (2008) reported that K deficiency had a negative impact on the grain yield of soybean, wheat, and maize. They suggested that crop productivity can be improved by increased application of K or by enhancing the efficient use of applied K via adopting 4R principles (Johnston and Bruulsema 2014; Murrell et al. 2021). One

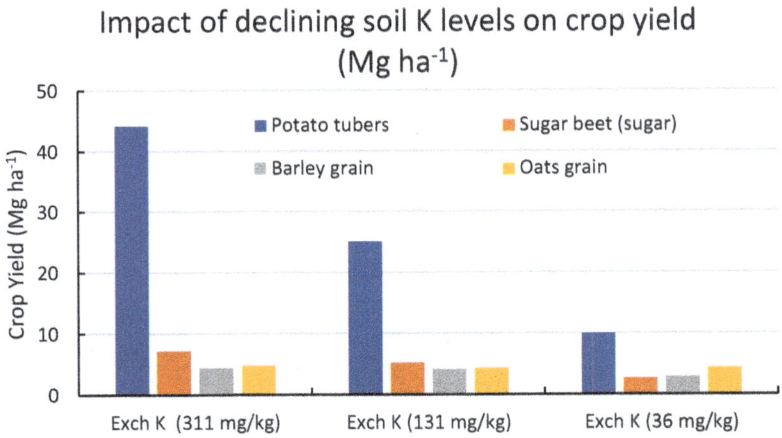

Fig. 4.6 Soil exch K levels on crop yield. (Source: modified from PDA 2019)

classical example of declining soil K levels having a significant impact on crop yield based on a long-term study at Rothamsted is given below (source: https://www.pda.org.uk/pda_leaflets/8-introduction-to-potash-use/potassium-in-the-soil/) (Fig. 4.6). Similarly, Ultisols, Oxisols, and Alfisols are generally low in total K, exchangeable K, and plant available K, and therefore, these soils require K application for sustainable crop productivity.

4.2.1.6 Exchangeable Ca and Mg

Calcium (Ca) and magnesium (Mg) are essential 'secondary nutrients', and their requirements stand in between macronutrients and micronutrients for crop growth. Depletion of exchangeable Ca and Mg occurs mainly due to leaching but also due to crop residue removal (Table 3.4), especially in acidic Alfisols, Ultisols, and Oxisols when these are cultivated for cropping. The reduction of organic matter content in these soils seriously reduces their cation exchange capacity and increases their zero-point charge (Gillman 1984) so that Ca and Mg are readily leached from their soil root zones. The crops grown on these soils, only after 1–10 years of cropping, exhibit not only Al toxicity but also Ca (and Mg) deficiency. In calcareous Vertisols, changes in the amounts of exchangeable Ca upon cultivation are generally negligible, although exchangeable Mg and Na may increase with cultivation (Dalal and Mayer 1986a; Standley et al. 1990; Kopittke and Menzies 2007). Stevens et al. (2005) and Kopittke and Menzies (2007) recommended an ideal ratio of Ca and Mg for attaining optimum crop yields.

4.2.1.7 DTPA (Diethylenetriaminepentaacetic Acid) Extractable Cu, Zn, and Mn

DTPA extractable Cu, Zn, and Mn have been used as indicators of plant available Cu, Zn, and Mn in calcareous soils (Hodgson 1963; Lindsay and Norvell 1978; Bageria et al. 2002) since these nutrients play a vital role in plant growth and metabolism. Deficiency or excess amount of these nutrients severely impacts crop growth and yield parameters (Bageria et al. 2002). Micronutrient availability and distribution are greatly influenced by changes in soil pH, organic matter, clay content, and hydro-thermal regime (pedo-climate) (Lindsay 1991; Brady 1995). Although the range of critical levels of Cu, Zn, and Mn varies, the mean levels have been shown to be 0.8 mg Cu kg^{-1} (range, 0.12–2.5 mg Cu kg^{-1}), 0.8 mg Zn kg^{-1} (0.25–2 mg Zn kg^{-1}), and 1.4 mg Mn kg^{-1} (1–2 mg Mn kg^{-1}). In Chromusterts, grey, brown, and red clays of the western Downs, Queensland, Grundon and Best (1982) found the critical level of DTPA extractable Cu of 0.4 mg Cu kg^{-1} soil, although in Alfisols, no crop response to added Cu was obtained even at 0.2 mg Cu kg^{-1} soil. Critical levels of DTPA-extractable Zn in Vertisols are pH-dependent (Haddad and Weir 1985), although in general, a concentration of 0.3–0.33 mg Zn kg^{-1} appears to separate Zn-responsive from Zn-non-responsive soils (Radjagukgauk et al. 1980; Dang et al. 1992). Katyal and Sharma (1991) studied 57 benchmark soils of India representing eight orders, namely, Entisols, Inceptisols, Aridisols, Vertisols, Alfisols, Ultisols, Oxisols, and Mollisols in India. Among micronutrients, Zn was found to be deficient in 47 out of 57 samples analysed, which suggested that parent material profoundly influences the micronutrient compared to soil moisture regime. Similarly, the availability of micronutrients (Fe, Zn, Mn, Cu) decreased as the soils tend to become aridic since these nutrients are relatively less mobile. Shukla et al. (2014) reported that the deficiency of micronutrients in soil was in the order of Zn, 36.5%, followed by B, 23.4%; Fe, 12.8%; Mn, 7.1%; and Cu, 4.2%, in India.

Decreases in DTPA extractable Mn, Cu, and Zn upon cultivation have been observed both in Vertisols and Alfisols. For example, Dalal and Mayer (1986a) found that the decrease in DTPA extractable Mn varied from 25% in Alfisols to as much as 70% in some Vertisols (Cecilvale clay), a decline twice as much as that of total organic C but similar to that of organic C in the light fraction (fPOM). In Waco clay, the rates of decrease in DTPA-extractable Cu and Zn were much lower than those in DTPA-extractable Mn (Table 3.8). In spite of the decreases in DTPA-extractable Mn, Cu, and Zn upon cultivation for 70 years, the Waco clay still contained above the critical concentrations of Mn, Cu, and Zn at the steady state (Table 3.8). Thus, only the sensitive crops (Haddad and Weir 1985), much longer periods of cultivation, unfavourable cultural practices (Thompson 1987), or diseases are likely to provide yield responses to the application of these micronutrients in Waco clay, although Dang et al. (1992) observed that Zn responses to wheat can occur at much higher concentration of DTPA extractable Zn than 0.3 mg Zn kg^{-1} soil if the soil contains substantial amounts of $CaCO_3$ content.

Some Vertisols are likely to be responsive to applied Zn and Cu (Grundon and Best 1982) even in their early periods of cultivation, especially if the soils are severely eroded. Severe micronutrient deficiencies also occur in some acidic Alfisols, Ultisols, and Oxisols (Sanchez 1987; Katyal and Sharma 1991; Shukla et al. 2014). Whereas in Vertisols, micronutrient availability to crops is constrained by the clay content, soil pH, and $CaCO_3$ content; however, in the acidic Alfisols, Ultisols, and Oxisols, these micronutrients are primarily lost by leaching because of low cation retention capacity. Thus, crop management practices that enhance arbuscular mycorrhiza fungi (AMF) would be much more effective in Vertisols than in Ultisols and Oxisols to increase micronutrient availability to crops. In the latter, SOM management and micronutrient fertiliser application would be the most appropriate means to alleviate micronutrient limitation to crop production (Fageria et al. 2002).

4.2.1.8 Consequences of Nutrient Element Loss from Soil

The nutrient depletion from the soil is a natural consequence of the cultivation of initially fertile soil for cereal grain cropping if the nutrients removed from the produce are not replenished (Dalal et al. 1997; Dalal 1998a, 1998b). In addition to the loss in economic produce, large losses of available soil nutrients can occur in a number of ways, such as through soil erosion and runoff, crop residue removal, burning, leaching, volatilisation and denitrification of N, reduced biological availability, increased fixation of less mobile elements (inorganic P, Cu, Zn, and Mn), and reduced nutrient use efficiency due to adverse soil conditions for plant growth (structural degradation, acidification, salinisation, sodification, soil-borne diseases). Consequently, grain yields decline, and grain quality deteriorates with increasing periods of cultivation if fertility restorative practices are not used (Lal 2015a, 2015b; Mosier et al. 2021). Concomitantly, the cost of production for each Mg (tonne) of grain produced increases sharply, thus reducing the gross profits and limiting the capacity to purchase necessary inputs, such as fertilisers and adoption of fertility restorative practices (Campell et al. 2004, 2007; Mosier et al. 2021). Therefore, soil nutrient depletion due to continuous cultivation and cereal grain cropping eventually becomes a constraint to the sustainability of agricultural production, degrades the resource base, and becomes economically unviable and ecologically unsafe (Dickinson 2003). A reduction in soil fertility by not replacing nutrients degrades the resource base, and hence, it is unsustainable (Dalal et al. 1997, Dalal 1998a, 1998b, Lal 2023).

4.3 Decline in Nutrient Bioavailability

It was mentioned earlier (Sects. 4.1 and 4.2) that due to the processes of soil fertility degradation, the loss of nutrient-supplying capacity, for example, for N, could even be proportionally greater than the loss of total organic matter. That is, the bioavailability of nutrient elements may decline faster than the total amounts in fertility-degraded soils, with concomitant diminution of the amount of nutrients available for plant biomass production (Campbell et al. 2007; Osman 2014).

There are a number of methods available to estimate the nutrient bioavailability in soil (Stevenson 1986; Norwell and Lindsay 1972), usually in the laboratory and glasshouse under constant temperature and moisture conditions. Since most of these methods are empirical in nature, an estimate of relative rather than absolute nutrient bioavailability of soil is obtained. Agents' contributions to nutrient bioavailability are briefly discussed.

4.3.1 Soil Enzymes

Nutrient bioavailability is controlled by bio-geo-chemical processes, which are mediated by enzymes originating from soil organisms and plants (Dick and Knedler 2005; Marschner and Rengel 2007). Thus, it was expected that enzymic activity in the soil would determine total soil biological activity and, consequently, that it would provide a 'fertility index' of soils useable for practical purposes in agriculture (Skujins 1978; Coleman et al. 1983; Kotroczo et al. 2014), such as an indicator of soil fertility degradation. Unfortunately, at the current level of understanding of soil enzymes, enzymic activity does not correlate well with soil and crop productivity (Bloem et al. 2005). This is partly because the total activity of an enzyme in the soil is comprised of activities associated with different soil components, for example, enzymes within the cytoplasm of proliferating microbial, animal, and plant cells; enzymes attached to the outer surfaces of viable cells, dead cells, and cell debris; enzymes secreted by proliferating cells, released on cell lysis; and abiontic enzymes which become associated with the organic matter due to adsorption, entrapment, or co-polymerisation during the formation of SOM (Burns 1982). It has not yet been possible to determine enzymic activities separately for each component in soil and even to identify the key enzymic activities that would indicate soil fertility degradation. Only a brief mention, therefore, is made of a few prominent enzymes studied in soil and indicates, wherever possible, the effect of continuous cultivation and cropping on enzymic activities in soil. For earlier work on soil enzymes, refer to Kiss et al. (1975), Burns (2013), and Nannipieri et al. (2018).

4.3.1.1 Dehydrogenases

These enzymes are considered to exist in the soil in intact microbial, animal, and plant cells, and thus, dehydrogenase activities generally reflect microbial respiration (Frankenberger Jr and Dick 1983) but not the number of microorganisms in the soil (Skujins 1973; Shaw and Burns 2006). Since dehydrogenase activity is generally related to the SOM concentration (Rao and Ghai 1985), it declines with depth along with decreasing SOM concentration (Musa and Mukhtar 1969). Also, the dehydrogenase activity in soil should decline with the loss of organic matter in cultivated soils. In an Alfisol, Rao and Ghai (1985) reported 42% lower dehydrogenase activity in an Alfisol, which was gypsum amended and cropped compared to that under grasses. However, Moore and Russell (1972) found that dehydrogenase activity in soil is unlikely to be a useful indicator of soil fertility status. Recently, it has been identified as a sensible indicator to understand different farm management practices such as tillage practices and crop residue and different nutrient management (Dick and Knedler 2005; Utobo and Tewari 2015; Blonska et al. 2020; Naorem et al. 2023).

4.3.1.2 Oxgenases

Both monooxygenases and dioxygenases are involved in the degradation of lignins, hydrocarbons, and herbicides and pesticides in soil. Since oxygenase activity, such as catechol oxidase activity, is not related to pH, moisture, or organic matter content of soil, oxygenase activity may be only a poor indicator of soil fertility degradation. However, oxygenases play vital roles in the metabolic conversion as well as degradative reactions of various natural compounds, namely, amino acids, lipids, hormones, and vitamins (Hayaishi 2013).

4.3.1.3 Cellulase

Pancholy and Rice (1973) showed that cellulase activity in soil was not correlated with total SOM, but it was related to the type of organic matter (cellulase content) added to the soil during the vegetative succession. Ross et al. (1982) observed that cellulase activities declined appreciably (74%), with a 54% decline in organic matter, when pasture soil was cultivated and cropped to maize for eight years. Deng and Tabatabai (1996) studied the effect of tillage and residue management on the activities of glycosidases, including cellulase, α- and β-glucosidases, and α- and β-galactosidases. They reported that the activities of these enzymes were significantly higher under NT/double mulch compared to other tillage and residue management treatments.

4.3.1.4 Proteinases and Peptidases

These are predominantly extracellular enzymes and may be immobilised in soil by entrapment within or bonded to the organic matter-clay matrix (McLaren 1975; Dick and Knedler 2005; Blonska et al. 2020). Due to steric hindrance, immobilised proteinases, which may be only active against low molecular weight substrates, are usually protected from biological degradation. This may be one of the reasons why the activities of these enzymes are not well correlated with either the total number of microorganisms or microbial activity in soils (Nannipieri et al. 1983; Loll and Bollag 1983; Liao et al. 2023).

4.3.1.5 Urease

Urease has been studied more intensively than other soil enzymes because it hydrolyses urea, a prominent nitrogen fertiliser used in agriculture (Dalal 1975; Mulvaney and Bremner 1978; Dharmakeerthi and Thenabadu 1996). Soil urease may be derived from plants, animals, and microorganisms and stabilised in soil matrix (organic matter-clay) as suggested for proteinases (McLaren 1975) because it is relatively stable in soil (Zantua and Bremner 1977). It has generally been found that urease activity is closely correlated with SOM concentration (Dalal 1975, Mulvaney and Bremner 1978; Rao and Ghai 1985), and therefore, urease activity is likely to be affected where soil fertility degradation is accompanied by the decline in SOM (Dharmakeerthi and Thenbadu 1996; Nannipieri et al. 1996).

For example, urease activities of a Vertisol declined considerably when it was cultivated for cereal cropping, although that of an Alfisol showed no significant decline upon cultivation (R.C. Dalal, unpublished data). Both soils lost organic matter on cultivation (Dalal and Mayer 1986a, 1986b). The inconsistent decline in urease activity with the decline in SOM may be due to soil pH effects, substrate, and temperature differences in these soils (Mulvaney and Bremner 1978; Dalal 1985), resulting in urease residing in various loci affording differential stability in soil.

Other researchers also reported that cropping system, tillage practices, nutrient management, and bio-climate had a significant influence on urease activity in soil (Velmourougane et al. 2013; Klose and Tabatabai 2000; Roscoe et al. 2000). Velmourougane et al. (2013) found that bioclimate significantly affect the average urease activity, and it was in decreasing order viz. sub-humid (moist) > sub-humid (dry) > semi-arid (dry) > arid. In another study, the level of fertilizer application (0 to 240 kg N ha^{-1}) did not affect the urease activity, soil organic matter, and microbial biomass N. Moreover, the microbial biomass N showed 97% and 69% of the variation in urease activity in the surface and sub-surface soil layers, respectively, in Oxisols of Brazil.

4.3.1.6 Phosphatases

Soil phosphatases are either phosphoric monoester hydrolases (such as phytase, nucleotidases, sugar phosphatases, and glycerophosphatase) or phosphoric mono- or diester hydrolases (such as nucleases and phospholipases). Soil phosphatases, by the catalysis of the hydrolysis of soil organic P into inorganic P, make it available to crops, and thus, they are responsible for organic P cycling in soil–plant systems and the environment (Dick and Tabatabai 1986, 1993; Nannipieri et al. 2011; Zhang et al. 2021; Liao et al. 2023).

Soil phosphatases exist in soil both as intracellular as well as extracellular enzymes, principally originating from microorganisms but also from soil animals and plants (Speir and Ross 1978; Dalal 1982; Dick and Knedler 2005). Because of their diverse origins, stability, and locations in soil, the amount of organic P mineralised is not always related to their phosphatase activity, since phosphatase activity probably becomes important only after the initial breakdown of soil organic matter, presumably catalysed by a myriad of microbial enzymes (Appiah and Thompson 1974; Nannipieri et al. 2011). In many Alfisols, Ultisols, and Oxisols, where organic P complexed with Fe and Al dominates, it is the relatively limited solubility of organic P that restricts its transformation and bioavailability to plants than the phosphatase activity per se. In Vertisols, where organic P complexed with Ca and clay dominates, its mineralisation appears to be similar to that of SOM (Sect. 4.2), although its relationship to soil phosphatase activity is not known. Ross et al. (1982) did report a substantial decline in phosphatase activity, along with organic matter, when a pasture soil was cultivated for maize for eight years. Dodor and Tabatabai (2003) studied the activities of phosphatases (acid phosphatase, alkaline phosphatase, and phosphodiesterase) under crop rotations and nitrogen fertilization in two long-term experiments (namely, Northeast Research Center (NERC) in Nashua and the Clarion Webster Research Center (CWRC) in Kanawha, Iowa, USA). Due to variations in SOM concentration, the enzyme activities are greatly higher in CWRC compared to the NERC site. Cropping rotation and N fertilization had a significant effect on phosphatase activities in soil.

4.3.1.7 Sulphatases

Soil sulphatases, such as arylsulphatase, catalyse the hydrolysis of organic S into inorganic S in soil and thus are essential in S availability and S cycling in the soil–

plant system and the environment (Tabatabai and Bremner 1970; Siwik-Ziomek et al. 2016). The sulphatase activities in Alfisols and Oxisols in Nigeria have been found to be significantly correlated with ester-sulphate S and total organic S (Cooper 1972) and in soils of low available S, with plant S uptake (Speir 1977).

Since sulphatase activity is significantly correlated with SOM, the decline of the latter in the soil on cultivation also results in the decline in sulphatase activity in the soil. Ross et al. (1982) found that sulphatase activity declined by 80% as compared to 50% for organic C loss, when a pasture soil (Ultisol) was cultivated for maize for eight years.

Ross et al. (1982), Frankenberger Jr and Dick (1983), and Dick and Tabatabai (1986, 1993) considered that soil enzymic activities may provide a more sensitive indicator than organic matter to changes in soil with cultural practices. It is too optimistic at present to relate the bioavailability of nutrients to the enzymic activity in soil because the latter is usually a combination of endocellular and abiontic (exocellular) enzymes with varying locations, stability, and environmental effects, such as pH, moisture, and temperature (Nannipieri et al. 2018). Moreover, most assays for soil enzymic activity involve simple substrates that are unlikely to exist in isolation in soil (Shaw and Burns 2006).

4.3.2 Adenosine 5′-Triphosphate and Adenylate Energy Charge

Adenosine 5′-triphosphate (ATP) is a useful indicator of biological activity in soil since it is a universal measure of total energy in biological systems (Eiland 1985; Inubushi et al. 1985; Dinesh et al. 2006). In living cells, coupling reactions of ATP to ADP hydrolysis typically shift the equilibrium ratio of products to reactants by about 10^8 times, which provides the driving force for nearly all the biochemical reactions in the cells. This also ensures that ATP is unlikely to survive long when the cells die or accumulate extracellularly in soil. Thus, it is a better indicator of biological activity in the living soil than many other enzymic activities, which may at least partly be the expression of biotic enzymes (Jenkinson and Oades 1979).

Oades and Jenkinson (1979) found that a Vertisol-carrying pasture contained more than 2.3 times ATP than that has been used for wheat cropping; corresponding organic C values varied by 1.8 times. In an Alfisol, however, corresponding values in ATP and SOC between these two cropping systems were essentially similar (under pasture about two times that under wheat-fallow). Ross et al. (1982) observed a much larger decrease in ATP (74%) than organic C (55%) when an Ultisol from under pasture was cultivated for eight years.

The ATP content in soil has been found to be closely correlated with soil microbial biomass (Jenkinson 1988) under 'standard' conditions in a number of soils but not in some soils (Domsch et al. 1979). For example, Fairbanks et al. (1984) found that the ATP measurements as indicators of bacterial biomass and activity were affected by moisture stress, nutrient variation, and protozoan predation and concluded that it was applicable only under low nutrient levels when neither bacterial numbers nor ATP levels changed. Thus, ATP may not provide a sensitive index of nutrient bioavailability or rate of nutrient cycling in soil.

It has been suggested that adenylate energy charge (AEC), ratio of (ATP + 0.5 ADP)/(ATP + ADP + AMP), measures the metabolic energy held in the adenine nucleotide pool and thus differentiates between actively growing and dividing cells (AEC, 0.8–0.95), resting cells (AEC, 0.5–0.75), and senescing cells (AEC < 0.5) (Atkinson 1977; Karl 1980). The AEC, therefore, may provide a sensitive measure of changes in nutrient cycling and bioavailability induced by cultural practices. Unfortunately, currently, the AEC measurements in soil are limited by the appropriate analytical techniques, and hence, no firm conclusions can be drawn (Martens 1985; Brookes et al. 1987). As the techniques improve for extraction and measurement of ATP, ADP, and AMP, their roles in delineating the amounts and proportion of soil microbial biomass into actively metabolising, stationary (dormant), or sensing microbial population—so that nutrient cycling and nutrient bioavailability can be quantified under different cultural practices—are becoming increasingly important. Also, soil moisture, temperature, and land use/cropping systems play significant roles in the ATP and adenylate energy charge of the MBC (Raubuch et al. 2002; Contin et al. 2000).

4.3.3 Soil Microbial Biomass

Soil microbial biomass largely comprises bacteria, archaea, and fungi and represents a living component of organic matter (Dalal 1998a, 1998b; Hoyle et al. 2006). Moreover, microbial biomass plays a vital nutrient pool in nutrient cycling, which is affected by several factors such as soil properties (soil pH, clay, organic carbon), climate, edaphic, farm management practices, and land use change (Lepcha and Devi 2020).

Although soil microbial biomass usually contains 1–4% organic C, 1–5% organic N, and 2–5% organic P, it is not only a source and sink of these plant nutrient elements which are more labile than the bulk of the same (Jenkinson and Ladd 1981), but also it is an agent of biochemical change in soil (Dalal 1998a, b). Furthermore, although there is no difference in the amounts of nutrients contained in microbial biomass between the soils in the temperate and tropical regions, the nutrient flux through microbial biomass and hence nutrient bioavailability is much faster in tropical soils.

Paul and Voroney (1980) estimated N flux through soil microbial biomass in English and Brazilian soils (Table 4.11). Microbial biomass N in both soils was

Table 4.11 Effect of continuous cultivation and cereal cropping on urease activities of Vertisols and Alfisols

	Urease activity (mg urea-N hydrolysed kg^{-1} soil h^{-1})			
	Vertisols (Waco)		Alfisols (Riverview)	
Soil depth (m)	Virgin	Cultivated (mean: 26 year)	Virgin	Cultivated (mean: 6 year)
0–0.1 m	72	52[a]	4.7	4.1
0.1–0.2 m	89	66[a]	3.5	3.4
0.2–0.3 m	84	68[a]	2.9	2.9

[a]Significantly lower at $P < 0.05$. Source: R.C. Dalal, unpublished data

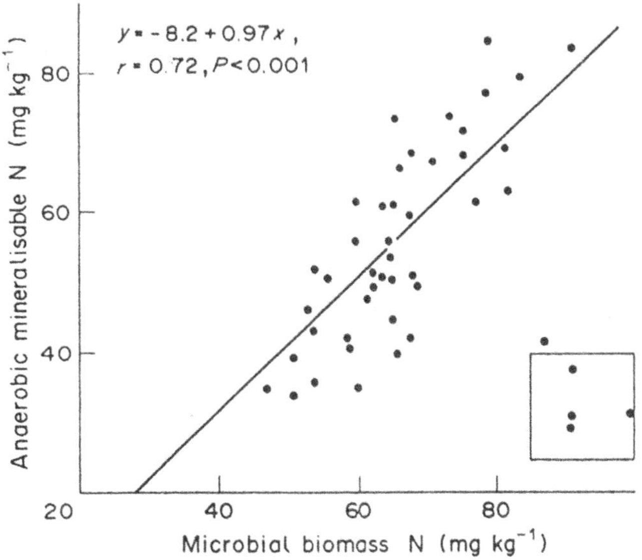

Fig. 4.7 Relationship between microbial biomass N and anaerobic mineralisable N. (Source: Dalal et al. 1991)

essentially similar (84–95 kg N ha^{-1}), although N flux through microbial biomass was ten times higher in the Brazilian soil (34 vs. 350 kg N ha^{-1} year^{-1}), mainly because of larger C inputs and faster C turnover. The consequences of accelerated biomass turnover and more gross mineralisation and immobilisation of nutrients are that following vegetation removal on land clearing, there may be a greater potential for rapid nutrient release and loss in tropical soils (Muller-Harvey et al. 1985) compared to that in temperate soils (Voroney and Paul 1984; Duxbury et al. 1989).

The difficulty with microbial biomass is that it does not measure nutrient bioavailability but a 'standing crop' (Jenkinson 1988) of nutrients, although it often correlates significantly with mineralisable N (Fig. 4.7). Improved techniques such as metagenomics and proteomics may be able to measure separately the function and components of soil microbial biomass and not only the nutrient bioavailability but also reflect accurately the extent of soil fertility depletion. There are three widely adopted methods for quantifying MBC: chloroform ($CHCl_3$) fumigation-incubation (CFI), chloroform ($CHC1_3$) fumigation-extraction (CFE), and substrate-induced respiration (SIR). Of late, molecular techniques in combination with fumigation have been prevalently used to identify/quantify microorganisms, namely, the contribution of fungi, bacteria, and archaea to the soil microbial community and function (Joergensen and Emmerling 2006; Joergensen et al. 2024).

Gross measurements of microbial biomass have frequently shown a rapid decline in the amounts of C and N contained in microbial biomass when virgin or pasture soils are cultivated (Srivastava and Singh 1989), especially when crop residues are not returned (Ayanaba et al. 1976; Dalal and Mayer 1987b) (Table 4.12). Consequently, both the nutrient pool of microbial biomass and nutrient flux through microbial biomass are depleted.

Table 4.12 Amounts of N contained in microbial biomass of soils under virgin/bush fallow and adjoining cultivated soils

Soil order	Microbial biomass N (mg kg^{-1} soil)		Decrease (%)
	Virgin/bush fallow	Cultivated	
Vertisol (Australia)[a]	52	42	19
Alfisol (Australia)[a]	22	12	45
Alfisol (Nigeria)[b]	93	49	47
Inceptisol (India)[c]	80	50	38

[a]Dalal and Mayer (1987b); [b]Calculated from Ayanaba et al. (1976); [c]Adapted from Srivastava and Singh (1989)

4.3.4 Mineralisable Nutrients

Soil fertility degradation is more closely related to the depletion of mineralisable nutrients than the total amounts when a soil under natural ecosystem is brought under cultivation and cropping (Ayanaba et al. 1976; Campbell and Biederbeck 1982; Dalal and Mayer 1987b; Dalal and Mayer 1990; Guo and Gifford 2002; Campbell 2007).

The bioavailability of nutrient elements in the soil is assessed by various means, for example, indirectly by soil physical, chemical, and biological characteristics and directly by measuring the amounts of nutrients in chemical extractions, soil incubations, ion-exchange membranes, plant uptake. Incubations of soil under temperature and moisture conditions prevailing in the field or standardised conditions in the laboratory and/or glasshouse are widely accepted techniques to measure bioavailable nutrients, especially N (Stanford 1982; Keeney 1982; Sahrawat 1983). Because of the almost universal limitation of N for crop production in fertility-degraded soils, examples are primarily derived from N studies.

Dalal and Mayer (1986a) found that total N and anaerobic mineralisable N declined in a Vertisol (Waco soil) by 35% and 55%, respectively, after a mean cultivation period of 26 years. In Alfisol (Riverview soil), after a mean cultivation period of only seven years, total N and anaerobic mineralisable N drastically declined by 28% and 51%, respectively. Thus, the decreases in anaerobic mineralisable N due to cropping exceeded those in total N, microbial biomass N, and even in N mineralisation potentials (N_o). For example, the rates of loss of total N, N_o, and anaerobic mineralisable N in Waco soil, cultivated for cereal cropping for up to 70 years, were 0.061 $year^{-1}$, 0.09 $year^{-1}$, and 0.11 $year^{-1}$, respectively (Fig. 4.7); for aerobic mineralisable N, it was even more rapid, 0.21 $year^{-1}$ (Dalal and Mayer 1990). In another Vertisol (Langlands-Logie), however, the rates of loss of N_o, anaerobic mineralisable N, and aerobic mineralisable and chemical indices, such as autoclavable N (Keeney 1982), were essentially similar to that of total N. Therefore, it is unlikely that any one of the current techniques of measuring the bioavailability of N would reflect similarly the mineralisability or lability of organic N in a wide range of soils (Stanford 1982; Sahrawat 1983; Dalal and Mayer 1990). Bioavailability of N, as measured by plant N uptake and dry matter yields, is better reflected by a combination of N measures, such as anaerobic mineralisable N and

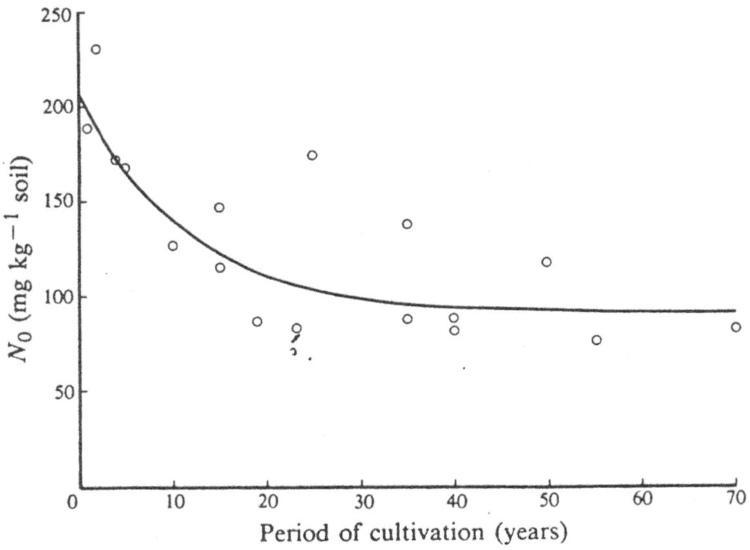

Fig. 4.8 Relationship between nitrogen mineralisation potentials (N_o) and period of cultivation in Waco soil (Typic Pellusterts). The curves are fitted according to the first-order Eq. 4.1d. (Source: Dalal and Mayer 1987a, 1987b)

profile NO_3-N at planting (Dalal and Mayer 1990), N_o and residual mineral N (Stanford 1982), and mineralised N during crop growth.

Chemical extraction methods are generally less sensitive to measure nutrient bioavailability and hence soil fertility changes, but for P, S, and micronutrient elements, these are almost exclusively used. This is because P, Cu, Zn, Mn, Fe, and to some extent S, once mineralised during incubation, are subjected to sorption, complexation, and precipitation reactions in the soil. The decreases in bicarbonate-extractable P, phosphate-extractable S, and DTPA-extractable Cu, Zn, and Mn in Vertisols and Alfisols subjected to continuous cultivation and cropping are referred to in Sect. 4.2.

4.3.4.1 Consequences of Reduced Nutrient Bioavailability

Indirect methods of estimating soil fertility degradation and nutrient bioavailability, such as soil enzymes, ATP, microbial respiration, and microbial biomass, are not directly related to crop productivity in spite of earlier expectations (Skujins 1978; Jenkinson and Ladd 1981). Potentially, mineralisable nutrient elements, estimated either by incubations or chemical extractions, predict better crop productivity and, hence, soil fertility degradation (Rashid et al. 2016). Dalal and Mayer (1990) found that dry matter production in a number of Vertisols and Alfisols declined as anaerobic mineralisable N was reduced with increasing periods of cultivation for continuous cereal cropping (Fig. 4.8). Stanford (1982) and Sahrawat (1983) listed a number of studies confirming the observations of Dalal and Mayer (1990).

4.4 Soil Physical Degradation

Soil physical degradation adversely affects the hydrologic, aerobic, thermal, and mechanical characteristics of soil and thus deleteriously affects crop productivity and sustainable land use (Lal 2009; Osman 2014; Wei et al. 2019). The interdependence among these soil characteristics has been demonstrated by Latey (1985); Lal (2020) Dexter and Czyz (2007).

4.4.1 Soil Hydrologic Characteristics

Decreased water infiltration rate on the cultivation of virgin and pasture soils has been frequently observed (Lal 1986; Cook 1988; Borrelli et al. 2018;), and it appears that in Alfisols, water infiltration decreases much faster than in Vertisols with the period of cultivation (Table 3.2). For example, the rate of loss in water infiltration in an Alfisol from Nigeria was about five times that of Vertisols from southern Queensland. The values of the rate of loss in water infiltration are somewhat similar to that of SOM loss (Tables 4.2 and 4.3) but much higher than the loss in water-stable aggregates >20 μm from the cultivated Alfisols and Vertisols.

The water infiltration rate decreases more rapidly in Ultisols than in Oxisols when these soils are cleared for cultivation; mechanised clearing appears to do the most damage. For example, Alegre (1985) found that the water infiltration rate decreased from 90×10^{-6} m s^{-1} before clearing to 4×10^{-6} m s^{-1} three months after clearing using a straight blade and only to 57×10^{-6} m s^{-1} using slash and burn method of land clearing.

Hydraulic conductivity decreases exponentially with increasing amounts of dispersed clay, which in turn was found to be linearly related to the increasing period of cultivation of Vertisols for cereal cropping (Cook et al. 1992). For example, as the dispersed clay increased from 40 g kg^{-1} soil to 120 g kg^{-1} soil, the hydraulic conductivity declined from 0.4×10^{-6} m s^{-1} to 0.05×10^{-6} m s^{-1}, thus emphasising the deleterious effect of aggregate breakdown on soil hydrologic characteristics (Sect. 3.6, Fig. 3.14). Similarly, Li et al. (2021a, b) reported that soil degradation (via soil erosion) had adversely impacted soil properties, i.e. decreased organic matter, saturated hydraulic conductivity (K_s), soil water content, capillary moisture capacity, field capacity, and increased bulk density, and reduced the ability of soils to retain water (Sarapatka et al. 2018; Borrelli et al. 2018).

The effects of reduced water infiltration rate and hydraulic conductivity with aggregate breakdown upon the cultivation of virgin and pasture soils are increased water loss through evaporation and runoff, soil erosion and nutrient loss through sediments, temporary waterlogging, and salt accumulation in the sub-soil. This leads to reduced plant available water in the soil profile and decreased available nutrients through lower beneficial biological activity (Sect. 3.6).

4.4.2 Soil Aeration

A corollary of the decline in water infiltration rate and hydraulic conductivity with increasing periods of cultivation, accompanied by aggregate breakdown, is that soil aeration is also adversely affected (Groenevelt et al. 1984; Anderson et al. 1990; Stepniewski et al. 1994 Blanco-Canqui et al. 2017). McGarry and Chan (1984) measured air-filled porosity in Vertisols 2, 5, and 10 days after an irrigation event. They used a 'critical' air-filled porosity of 0.1 m^3 m^{-3} to study the effect of the cultivation period for cotton (1 and 10 years) down to 0.5 m depth. They found that a Vertisol which had grown cotton for one year attained critical air-filled porosity throughout 0 to 0.5 m depth by day 5. However, a similar Vertisol which had grown cotton for 10 years had less than critical air-filled porosity below 0.2 m depth, even after 10 days. Furthermore, the air permeability even in the top 0 to 0.2 m layers was much lower in the latter (after 1-year cotton, 0–0.1 m and 0.1–0.2 m and 828 and 81 μm^2 and, after 10-year cotton, 105 and 5 μm^2, respectively). Mentges et al. (2016) studied the air-filled porosity and pore continuity in Oxiosls and Ultisols under a no-till system. They reported that air-filled porosity and pore continuity were reduced due to soil compaction. Soil deformation had an adverse effect on the function of the soil pore system and airflow. Moreover, soil properties such as soil texture, clay content, pore size distribution, tortuosity, and pore connectivity greatly influenced the oxygen diffusion rate (Neira et al. 2015).

Unfortunately, air-filled porosity is a static measure of the aeration status of soil. Also, it is not closely correlated with oxygen diffusion rate or oxygen flux density, a more pertinent measure of soil aeration, although air permeability provides an improved measure of soil aeration. In fact, both air-filled porosity and air permeability along with moisture characteristic curves can be used to estimate the tortuosity of the oxygen diffusion path in soil (Deeks et al. 2004; Troeh and Thompson 2005a, 2005b; Hillel 2007). Hodgson and MacLeod (1989) compared oxygen flux density, air-filled porosity, and bulk density as a measure of structural changes in a Vertisol. They found that oxygen flux density was the most sensitive of the three indices because it depended on air-filled porosity as well as the tortuosity and continuity of air-filled pores. That is

$$D/D_o = S \tag{4.4}$$

where D and D_o are diffusion coefficients of oxygen in the soil and in free air, respectively, S is air-filled porosity, and α is a measure of the tortuosity and continuity of air-filled pores, which vary with soil texture, soil structure, and moisture content (Currie 1970; Stepniewski et al. 1994; Hillel 2007). In another study, Blackwell et al. (1990) found that intrinsic permeability is the most sensitive parameter to structural change and pore organisation (*E*), which is largely governed by the dominant shapes of the macropores in Australian red-brown soils. Berisso et al. (2012) studied the continuous effects of sub-soil compaction on soil pore and gas transport in loamy soil (Luvisol), in Sweden. They found that persistent compaction (14 years with ~10 Mg wheel loads) reduced the air permeability and gas diffusivity in this soil.

The flux of O_2 into soils in response to the diffusion gradient created by respiration activities of roots and microorganisms is about 5 to 25 g m^{-2} day^{-1} in tropical and subtropical regions (MAT, 17–25 °C) (Currie 1970). Since the volume of oxygen contained in the soil profile to 0.3 m depth is only about 9.4 g m^{-2} at an air-filled porosity of 0.1 m^3 m^{-3}, it is only sufficient for less than two days of respiration activity. Since the gaseous and dissolved O_2 content of the soil profile is small, optimum oxygen flux density is essential for crop production (Hodgson and MacLeod 1989). The oxygen flux density is substantially reduced if excess water is present in the soil since the oxygen diffusion coefficient through a water film is about 10^{-4} times that in air, 2.4×10^{-9} m^2 s^{-1} and 2.1×10^{-5} m^2 s^{-1}, respectively.

It is apparent, therefore, that soil degradation that leads to reduced water infiltration rate and hydraulic conductivity would also result in poor oxygen flux density in soil when a wet event occurs. Unfortunately, information is lacking on oxygen diffusion rate or oxygen flux density in Alfisols, Oxisols, Ultisols, and Vertisols of the tropical and subtropical regions to evaluate the consequences of land management when virgin and pasture soils are brought under cultivation.

4.4.3 Soil Temperature

Soil temperature is affected by the amount of radiation the soil receives, its albedo, its moisture and air contents, and any factor that affects the rate of evaporation of water from the moist soil (Payne and Gregory 1988; Troeh and Thompson 2005a, 2005b; Hillel 2007).

Vegetation on virgin and pasture soils and mulches on the soil surface reduces the diurnal temperature fluctuations in soil. For example, Lal and Cummings (1979) observed a maximum temperature of only 28 °C in an Alfisol under the forest compared to 36 °C in an adjoining cleared Alfisol at 0.1 m depth; the maximum temperatures varied by 20–25 °C in the top 0.01 m depth. Similarly, mulches can reduce the maximum soil temperatures by 5 to 10 °C (Nangju et al. 1975; Bristow 1988). This is of considerable significance in semi-arid and sub-humid tropical and subtropical regions, whereby by reducing supra-optimal temperatures (< 30 °C), both seedling emergence and crop establishments can be improved considerably (Lal 1978; McCown et al. 1980, 1983). However, almost nothing is known about changes in soil temperatures as soil degradation occurs, although it may be surmised that with reducing organic matter and crop yields, the amount of crop residues returned to the soil would decrease, and hence, residue cover would also decrease. As a consequence, soil temperature fluctuations are likely to be greater with increasing state of degradation of the soil (Lal 2009; Osman 2014).

The rate of change of temperature (dT/dt) at a given depth depends on the difference between incoming and outgoing heat flux in the soil at that depth and on its thermal capacity, that is

$$dT/dt = Dq\, d^2T/dz^2 \tag{4.5}$$

where $Dq = kq/P_b c_p$ is the thermal diffusivity of soil and kq is the thermal conductivity of soil (soil air, 0.025 Wm^{-1} K^{-1}; soil water, 0.6 Wm^{-1} K^{-1}; soil solids, ~ 9 Wm^{-1} K^{-1}). P_b is the bulk density of soil, and c_p is its specific heat capacity (solids + water + air). Although kq increases with increasing moisture content in most non-swelling soils, it increases more rapidly than $P_b c_p$ in moderately dry soil, but less rapidly at higher moisture contents, so that Dq initially increases with increasing soil moisture content, reaches a maximum, and then decreases. However, in Vertisols, although kq increases initially with increasing soil water content, similar to Dq, it decreases at higher moisture content as soil solids move apart (Ross and Bridge 1987). Since soil water content affects both kq and cp and hence Dq in soil, the effect on dT/dt could be positive or negative. Furthermore, in Vertisols, P_b decreases as the soil water content increases due to swelling of the soil. As a result, after achieving its maximum value at ≤0.2 kg H_2O kg^{-1} soil, Dq decreases much faster with increasing moisture contents in Vertisols than in Alfisols, Oxisols, and Ultisols.

Soil degradation accompanied by the loss of SOM usually leads to increasing bulk density (P_b) of soil (see below), which affects both 'kq' as well as 'cp'. The 'kq' increases much faster than 'cp' at a given moisture content, so Dq may increase as soil bulk density increases. Thus, soil degradation may lead to not only increasing maximum temperatures but also greater diurnal temperature fluctuations up to a greater depth in the soil profile. Unfortunately, soil temperature variations experienced in different soils due to soil degradation have not been studied in detail. Since soil temperature considerably influences soil heat flux density (de Vries 1975; Fuchs 1986; Hillel 2007), soil water, especially soil biological activity and crop growth; and hence soil productivity, changes in soil temperatures as a result of soil degradation need to be studied in detail (Hillel 2007; Evett et al. 2012).

4.4.4 Soil Mechanical Characteristics

Soil mechanical resistance impairs seedling emergence and/or root growth and thus affects crop growth (Hillel 2007). Two measurements, penetration resistance and soil bulk density, have been widely used to quantify the mechanical resistance of the soil.

Generally, soil physical degradation has been accompanied by increasing resistance to cone penetrometers (Goedart 1983). For example, Goedart (1987) reported an increase in penetrometer resistance from 0.3×10^{-4} kg m^{-2} in a virgin condition to 12×10^{-4} kg m^{-2} at 0.12 m depth after 10 years of intensive cropping of an Oxisol in Brazil, an increase in mechanical resistance by 40 times. Lal (1989) measured an increase in penetrometer resistance from 25 kPa when an Alfisol was cleared of its secondary forest regrowth to 210 kPa after 4 years under maize-cowpea rotations in Nigeria.

However, the interpretation of penetrometer readings as a measure of soil mechanical resistance to seedling emergence and root growth is complicated by a number of interdependent soil characteristics, especially soil water content. Dexter (1987) proposed that relative root elongation (R/R max) can be measured as a function of soil water potential and penetrometer resistance, that is

$$R/R\ \max = \exp\left[-0.693\ (Qp/Q0.5) - \psi_o/\psi_{\max}\right] \tag{4.6}$$

where Qp is the penetrometer resistance, Q0.5 is the penetrometer resistance at relative root elongation of 0.5 (which is for most crops, 0.72–2.03 MPa), and ψ_o and $\psi_{\max}$ are the soil water potential and wilting point potential of crops, respectively. Although Eq. (4.6) takes into account soil water potential besides the penetrometer resistance, it does not account for the lubrication of the soil by plant root exudates and the ability of a plant root to explore and use the path of least resistance (Taylor and Klepper 1974; Taylor 1980).

Bulk density measures soil compaction directly and thus provides a direct measure of soil physical degradation, although its interpretation in terms of its effect on crop growth has similar limitations as that of penetrometer resistance (Batey 2009; Bengough et al. 2006). Moreover, bulk density is inadequate in measuring surface sealing and soil crusts (Nciizah and Wakindiki 2015a,b; Valentin and Bresson 2020). Further details are available elsewhere (Letey 1985; Raghavan et al. 1990; Mullins et al. 1990; Valentin and Bresson 2020).

In many soils, especially Alfisols, Entisols, and Inceptisols, soil degradation accompanied by the loss in SOM also leads to an increase in soil bulk density. Harte (1984) recorded an increase after 15 years of continuous cultivation and cropping in bulk density (0–0.2 m) from 1.30 Mg m^{-3} to 1.62 Mg m^{-3} (25% increase) in a loamy Alfisol, from 1.42 Mg m^{-3} to 1.62 Mg m^{-3} (14% increase) in a sandy Alfisol, and from 1.13 to 1.22 Mg m^{-3} (8% increase) in an Oxisol in northern New South Wales, Australia. In a cultivated sandy Alfisol, soil bulk density reached 1.71 Mg m^{-3} at 0.08 to 0.15 depth. Harte (1984) also found significant correlations between SOM concentration and bulk density in these soils, and thus, substantial increases in bulk density were associated with declines in SOM. In southern Queensland, Dalal and Mayer (1986a, 1986b, 1986c, 1986d, 1986e) measured increases of 13% to 28% in bulk density of Vertisols and Alfisols (0–0.1 m) on the cultivation of the virgin soils for periods of up to 70 years. Similar results were obtained by Cook et al. (1992) for two Vertisols at 0–0.1 m as well as 0.1–0.2 m depths. Anderson et al. (1990) reported that 100 years of cultivation coupled with the annual addition of manure at 13.5 Mg ha^{-1} not only reduced bulk density by 0.12 Mg m^{-3} compared to the unfertilized treatments but also increased saturated hydraulic conductivity by nine times. However, the soil and crop management had an inconsequential effect on bulk density and pore-size distribution.

The kinetics of changes in bulk density of two Vertisols and one Alfisol on cultivation for cereal cropping are given in Table 4.13. Whereas in a Vertisol, Waco clay (72% clay content), the rates of increase in bulk density were small, especially at 0.1–0.2 m depth (0.02 $year^{-1}$), in Riverview sandy loam (Alfisol), the rates of

Table 4.13 Initial (ρi) and equilibrium (ρe) bulk densities and rate of increase in bulk density at two depths of Vertisols and Alfisols cultivated for cereal cropping[a]

Soil series	Periods of cultivation (year)	Bulk density ($Mg\ m^{-3}$)				r ($year^{-1}$)	
		0–0.1 m		0.1–0.2 m		0–0.1 m	0.1–0.2 m
		ρi	ρe	ρi	ρe		
Waco clay[b]	0–70	0.86	1.09	0.97	1.20	0.06	0.02
Billa Billa loamy clay[c]	0–25	0.99	1.39	1.10	1.44	0.09	0.11
Riverview sandy loam[d]	0–20	1.25	1.47	1.24	1.42	0.62	0.65

[a]R.C. Dalal, unpublished data. Equation used: $\rho t = \rho e + (\rho i - \rho e) \exp (rt)$, where t is period of cultivation in years. Measured bulk densities of Vertisols were calculated at field capacity soil water content

[b]Typic Pellusterts, [c]Typic Chromusterts, [d]Rhodic Paleustalfs

increase in bulk density (0–0.1 m and 0.1–0.2 m) were more than ten times as fast as that in the former. When a decline in organic matter is accelerated by keeping the soil under bare fallow, the rate of increase in bulk density of an Alfisol (0.05–0.15 m) in Thailand was found to be even higher (1.26 $year^{-1}$ for the first two years after forest removal than that reported in Table 4.13 for an Alfisol. The increase in bulk density following the loss in organic matter is much greater in soils where organic matter is prominent in aggregate formation and stabilisation (Entisols, Inceptisols, Ultisols, and sandy Alfisols) than where it is less so (Vertisols, where clay matrix and shrink-swell characteristics predominate, and Oxisols, where oxides of Fe and Al predominate).

Crop yields are adversely affected by the increasing bulk density of soil ($>1.6\ Mg\ m^{-3}$). In Alfisols from Senegal, sorghum yields were inversely related to the soil bulk density, whereas peanut pod yield declined exponentially with increasing bulk density (El-Saif et al. 1987) (Fig. 4.2). Since the root mass of these crops was similarly affected by soil bulk density, mechanical resistance to the root growth may have been mainly responsible for reduced crop yields; this would have restricted the access to plant available water in the soil profile (Bengough and Young 1993; Bengough 2003; Bengough et al. 2006; Batey 2009).

4.4.5 Interrelationship Among Various Measures of Soil Physical Degradation

Of the four measures of soil physical degradation that may directly reduce crop growth in the semi-arid, tropical and subtropical regions, soil hydrologic aspect, especially soil water, is the dominant controlling factor which also affects the other three: aeration, temperature, and mechanical resistance of soil (Jayaraman et al. 2018a, b, c, d, 2019). Letey (1985, 1991) generalised the relationships between soil water content and aeration and mechanical resistance. He proposed the non-limiting water range (NLWR) which may be equal to or less than PAWC depending upon aeration and/or mechanical resistance, particularly in soils that have high bulk density and exclusively small pores. That is, with an increase in bulk density and/or aggregate breakdown as soil degrades, the NLWR becomes narrower.

Unfortunately, there is a scarcity of information on the rate of change of NLWR with soil degradation and even its relationship with aeration and mechanical resistance of soil. Almost nothing is known about the relationship between NLWR and PAWC. The response of plants to variation in water content must be considered one of degree—inside the range (Letey 1985), growth is least limited, while outside of the range, growth is most limited. Thus, the term *least limiting water range* (LLWR) is a better parameter rather than a non-limiting water range to describe the concept introduced by Letey (1985). LLWR is the range of soil water content within which plant growth is least limited by water potential, aeration, and mechanical resistance. The integrated effect of matric potential, air-filled porosity, and soil strength on plant growth was characterized by a single parameter, i.e. least limiting water range (LLWR). The LLWR defines *a range in soil water content within which plant growth is least likely to be limited by the availability of water and air in soil and soil strength* (da Silva et al. 1994).

Therefore, it is essential for plant production in water-limiting environments such as in semi-arid and sub-humid tropical and subtropical regions to understand the dynamics of NLWR or LLWR and PAWC in soil with continuous cultivation and cropping and their relationships with soil physical, chemical, and biological aspects of land degradation.

4.5 Soil Biological Degradation

Soils contain a wide range of biological forms besides plant roots (Anton 2017; Guerra et al. 2020) (Fig. 4.9). These vary in size from the submicroscopic virus and bacteriophage particles, through microflora, <100 μm body width (bacteria, archaea,

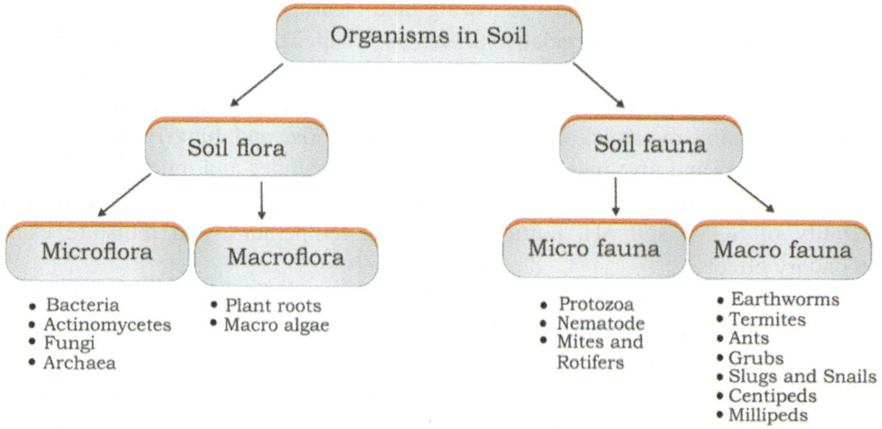

Fig. 4.9 Classification of different soil organisms. (Source: authors developed from many literature)

actinomycetes, fungi, and algae); microfauna, <100 μ (nematodes and protozoa); mesofauna, <2 mm (collembola, acari, and Enchytraeidae); to macrofauna, > 2 mm (Lumbicidae, Mollusca, Hymenoptera, Isoptera, Isopoda, Diplopoda, and Chilopoda and numerous other soil-dwelling insects). The soil organisms mediate biological processes such as SOM turnover (saprophytic) and nutrient cycling, nitrogen fixation and nutrient scavenging (symbiotic), and soil-borne plant diseases (parasitic). They also substantially modify the soil chemical environment (pH, oxidation, reduction, and complexation) and physical environment (aggregation, soil pores, and channels) (Hillel 2007).

Soil degradation following land clearing and cultivation for cropping alters the numbers as well as the composition of soil organisms (Wallenstein 2017; Mosier et al. 2021). In the long term, the decrease in numbers of saprophytic and symbiotic soil organisms, increase in parasitic organisms, and decrease in biodiversity lead to reduced soil productivity and hence soil degradation. Soil biological degradation and soil faunal degradation were reviewed by Sims (1990) and Curry and Good (1992). Examples are given in this section of changes in saprophytic, symbiotic, and parasitic organisms and their predators following land clearing and cultivation for cropping, especially monoculture in the semi-arid and sub-humid tropical and subtropical environments.

4.5.1 Decrease in Beneficial Soil Microflora and Soil fauna

4.5.1.1 Saprophytes and Detritivores

Land clearing and cultivation lead to a reduced amount of organic matter in soil and a reduced input of fresh organic materials from cropping results in lower SOM stock (Sect. 4.1). Consequently, saprophytic organisms, which derive their energy from the decomposition of SOM, also decline along with their predators such as bacteriophagus protozoa and nematodes. Although such trophic interactions in soil are essential for organic matter turnover, nutrient cycling, and soil structure, quantitative changes in soil organisms following land clearing and cultivation are rarely studied except where soil degradative processes (Chap. 2) could be clearly identified.

The trophic interactions shown in Fig. 4.3 are explained primarily in terms of predator-prey interactions (Ingham et al. 1986) which explain the seasonal changes in the population dynamics of prey (archaea, bacteria, and fungi) and predators (bacteriophages and mycophagous protozoa and nematodes and mites). It is generally agreed that long-term population changes are related primarily to the quality and quantity of SOM; soil pH, nutrients, and soil temperature and moisture exert secondary effects.

Swift et al. (1980) observed that the continuous cropping of an Alfisol in sub-humid tropical Nigeria disrupted SOM decomposition and nutrient cycling processes by (1) reducing the amount and diversity of litter input; (2) decreasing population size, species composition, and diversity of saprophytes and detritivores; and (3) reducing the time interval between litter inputs and changing the site of decomposition from the soil surface to within the soil. Of these decomposers, bacteria account for almost 70% of SOM turnover and N cycling in soil since they contain diverse metabolic capabilities and utilise many sources of energy and C in soil. The soil macrofauna, earthworms, termites, and ants account for a further 10–15% in forest (Lavelle 1988) and grassland soils (Dash and Patra 1977) but much less in cultivated soils.

The two major beneficial associations between soil microflora and plants are nitrogen fixation (*Rhizobium* and *Bradyrhizobium* associations with legumes and that of *Frankia* with non-legumes) and mycorrhizal associations, especially arbuscular mycorrhizal fungi (AMF) association with crop plants.

4.5.1.2 Rhizobium and Bradyrhizobium

Herridge and Bergersen (1988) provided estimates of N_2 fixation in tropical legume crops; the amounts of N_2 fixed varied from 0 to 236 kg N ha^{-1} for soybean, 47 to 120 kg N ha^{-1} for cowpea, 94 to 222 kg N ha^{-1} for groundnut (peanut), and 13–69 kg N ha^{-1} for pigeon pea. These wide variations in the amounts of N_2 fixation may be due to a number of factors such as soil acidity, soil sodicity, plant cultivars, cultural practices, nutrient deficiency such as P and Mo, excessive nitrates in the soil, and inappropriate and insufficient *Rhizobium*/*Bradyrhizobium* bacteria.

Soil degradative processes, such as soil erosion which results in SOM loss; nutrient depletion; increased soil acidity in Ultisols, Oxisols, and acidic Alfisols; increased sodicity in Vertisols; and aggregate breakdown, can reduce the survival and infectivity of root nodules. Habte and El-Swaify (1988) observed up to a 100-fold decrease in the numbers of indigenous rhizobia in an Oxisol subjected to simulated erosion in excess of 0.15 m of soil removal. This was closely related to the increase in soil acidity; soil pH declined from 5.6 to 5.0 at >0.15 m depths.

4.5.1.3 Arbuscular Mycorrhizae Fungi (AMF)

The AMF forms an association with the roots of most plants (Linderman et al. 1997). These associations primarily enhance the uptake of relatively slowly mobile nutrients such as P, Zn, Cu, and Fe. Since AMF is propagated primarily through plant roots and multiplies in association with roots, the relatively low intensity of cropping following land clearing and cultivation leads to the reduced level of AMF population and decreases AMF infection of crops following clean fallowing (Table 4.14).

Table 4.14 Effect of duration of clean fallowing on arbuscular mycorrhizal fungi (AMF) infection of roots of the following crop[a]

Soil	Crop	Period of clean fallowing (month)	Root length infected with AM	
			%	m/plant
Typic Pellusert	Chickpea	6	72.7	–
		14	17.5	–
Typic Chromustert	Sorghum	6	77.2	38.9
		23	25.1	4.5

[a]Extracted from Thompson (1987)

Soil acidification following land clearing and continuous cultivation of Ultisols, Oxisols, and acidic Alfisols (Sect. 3.4) can lead to decreased root infection by AMF. Zhao (1992) demonstrated a linear decline in cotton total root length as well as the percentage of root infected by *Glomus fasciculatum* (Fig. 4.3). Below pH 5.5, less than 25% of the cotton root could be infected by *G. fasciculatum*, whereas greater than 50% of the root length was infected at pH 6.5.

Soil erosion accompanying the removal of topsoil en masse substantially reduces AMF spores as well as propagules so that the AMF-dependent crops grown in severely eroded soils could be adversely affected in their P and Zn uptake and hence yield if they are poorly colonised by AMF.

4.5.1.4 Pathogen- and Disease-Suppressive Soil Organisms

Pathogen suppression occurs when the pathogen is growing saprophytically or surviving in the soil and disease suppression describes suppression of the pathogen growing parasitically in the host (Hornby 1983; Jayaraman et al. 2021). Most species of *Trichoderma* and *Gliocladium* produce toxic metabolites such as gliotoxin, gliorium, trichodermin, and viridian in pure culture (Papavizas 1985), including antibiotics, although these have not been successfully isolated from soil in significant amounts. These species suppress pathogens and diseases by a number of mechanisms, including antibiosis, mycoparasitism, lysis, and competition. The population dynamics of *Trichoderma* and *Gliocladium* are closely correlated with the amount of organic matter in soil; soil pH and available nutrients, especially Fe and Mn, also affect their populations. It is, therefore, expected that soil degradation accompanying SOM loss would result in reduced populations of *Trichoderma* and *Gliocladium*. Unfortunately, lack of information precludes arriving at a definitive relationship between organic matter content and populations of *Trichoderma* and *Gliocladium*. However, Wells et al. (1972) reported that the field control of *Sclerotium rolfsii* on tomato plants by *T. harzianum* required large amounts of organic matter additions (4.2 Mg ha^{-1}), thus indicating the importance of SOM concentration in maintaining the population of *Trichoderma*.

Much less is known about the population dynamics of other potential mycoparasites such as *Taloromyces flavus*, *Sporidesmium sclerotivorum*, *Laetisaria arvalis*, *Pythium nunn* (Adams 1990), *Pseudomonas* sp., *Flavobacterium oderatum*, and *Bacillus* sp. (Hebbar *et al.* 1991). This also holds for the potential phytonematophagous parasites such as fungi (e.g. *Arthrobotrys* sp., *Paecilomycetes* sp.), bacteria (e.g. *Pasteuria* sp.), and nematophagous mites (e.g. *Lasioseius* sp.) (Sayre and Walter 1991).

4.5.1.5 Earthworms, Termites, and Ants

Earthworms, termites, and ants comminute and disperse plant residues; thus, they participate in organic matter turnover and, through their burrowing activities and reworking of soil, also provide macropores for water movement and aeration (Linden et al. 1994; Lavelle et al. 2001, 2004). Also, they participate in nutrient cycling by harbouring microflora and microfauna in their digestive tracts. Earthworms are the dominant group in sub-humid tropics, and termites and ants, especially the latter (de Bruyn 1993), are the dominant groups in the semi-arid regions. Soil degradation in the tropical and subtropical regions following land clearing and cultivation is associated with reduced numbers and activity of earthworms, termites, and ants.

Earthworm populations generally decrease with the decreasing mean annual rainfall (Lavelle 1983). Their activity depends upon soil type, soil temperature and moisture, organic residue availability, and cultural practices such as tillage (Lee and Wood 1971; Lee 1985; Lee and Foster 1991; Lal 1988; Wildermuth et al. 1993, 1997).

Initial forest clearing and cultivation drastically reduced the number of earthworms in an Alfisol in Nigeria (Critchley et al. 1979); bush fallowing increased the numbers of earthworms (37 m^{-2}), but the cultivation of soil reduced to a very low population (6 m^{-2}). Even larger differences were observed in a Vertisol at 0-0.1 m depth, where the earthworm population in soil under 30 months' pasture was 440 m^{-2} compared to that of 11 m^{-2} under continuous cultivation for cereal cropping (R.C. Dalal, unpublished data).

Cultivation reduces not only the number of earthworms in soil but also changes the species composition (Lee 1985; Lee and Wood 1971; Lee and Foster 1991). For example, Cook et al. (1980) observed that the surface casting activity of *Hyperiodrilus* sp. decreased and that of *Eudrilus* sp. increased when the soil (Alfisol) under bush fallow was brought under cultivation, although total worm casts decreased with increasing periods of cultivation. Since wormcasts contained two to three times more organic C and N, two to five times more Ca, three times more K, and three to five times more P than the bulk soil under cultivation (Cook et al. 1980; De Vlees-chauwer and Lal 1981; Krishnamoorthy 1989), reduced wormcasts in continuous cultivated Alfisol would lead to soil degradation, and in the long term, it can lead to reduced crop productivity.

Since the number of earthworm casts is related to earthworm channels which contribute to macroporosity and so influence water infiltration and aeration (Lee and Pankhurst 1992), their reduced numbers in soil under continuous cultivation would result in a decreased rate of water infiltration (Lal 1988). This is of considerable importance in regions that receive high rainfall intensity storms where lack of earthworm channels can substantially increase runoff and soil erosion.

Both ants and termites comminute plant residues (Fig. 4.3) and thus participate in plant decomposition (Gupta et al. 1981). Soil turnover rate by termites (*Isoptera*) is usually <5 Mg ha^{-1} $year^{-1}$, while that by ants (*Hymenoptera*) may be twice as much (Lal 1988; De Bruyn and Conacher 1990). Since these are social insects, their populations are highly aggregated, and their effects on soil properties are concentrated; about 1–2% of the total surface area is affected by their activity (de Bruyn and Conacher 1990).

The organic matter turnover within termite mounds has been reported to be higher than in the adjacent bulk soil. Holt (1987) estimated that *Amitermes laurensis* mounds on a virgin Alfisol in a semi-arid region contributed 4–10% of total organic C mineralised from soil. Generally, organic matter, clay, and exchangeable Ca contents of mounds exceed that of the bulk soil (Lee 1985; Holt and Lepage

2000). Although water infiltration rates of mounds are lower than that of the bulk soil, that of galleries and channels could be much higher (Wielmaker 1984; Lal 1988), and thus, these can reduce surface runoff and soil erosion.

Unfortunately, studies on the effects of cultivation on the population and activity of termites are very few, and therefore, no definite conclusions can be drawn on the effect of drastic reduction of termites on soil degradation, both of which occur when virgin soil is brought under cultivation.

It has been observed that ant mounds in contrast with termite mounds are usually not enriched with clay, although the former may still be enriched with organic matter and P compared to the bulk soil (de Bruyn and Conacher 1990). Again, there is hardly any information on the effects of cultivation following land clearing on the population and activity of ants. In an Ultisol, de Bruyn (1993) recorded a maximum of 17 ants m^{-2} in soil under cultivation compared to 114 ants m^{-2} in the adjacent virgin soil. Majer (1983) suggested that ants could be used as sensitive bioindicators of degraded and disturbed soils.

However, the significance of both ants and termites on crop productivity is determined by a dynamic balance between their beneficial effects on some soil characteristics, especially in poorly productive Alfisols and Ultisols in semi-arid regions, and their pest status for crops, which fluctuates with seasonal conditions.

4.5.2 Increase in Harmful Soil Microflora and Soil Fauna

Soil-borne and soil-inhabiting harmful soil microflora and microfauna (primarily root-infecting bacteria and fungi, protozoa, and nematodes) and root-attacking soil fauna (mites and other insects) cause damage to plants and hence reduce crop yields (Wildermuth et al. 1992).

Land clearing and cultivation for continuous cropping drastically change the soil environment and thus alter the number and composition of soil microflora and soil fauna. Soil degradation, usually accompanied by organic matter loss, nutrient depletion, aggregate breakdown, and changes in water and aeration status and pH (Dalal and Mayer 1986a), results in reduced biodiversity. This may lead to an increase in soil-borne diseases and damage to plants by soil mesofauna (e.g. mites) and macrofauna (e.g. insects).

4.5.2.1 Soil-Borne Diseases

Soil-borne diseases form only a small fraction of total number of plant diseases. Kommedahl and Windels (1979a,b) listed 50,000 plant parasitic and nonparasitic diseases in the USA. Although only 6.7% were classified as root diseases, they cause considerable yield damage and hence result in substantial economic loss in the sub-humid and semi-arid subtropical and tropical regions.

Fusarium diseases

Plant diseases caused by soil-borne *Fusarium* spp. in some economically significant crops in tropical and subtropical regions include *F. acuminatum* (*Medicago sativa*, maize), *F. culmorum* (cereal root rot), *F. equiseti* (*Medicago sativa*, maize, wheat), *F. graminearum* (crown rot of wheat), *F. moniliforme* (stalk rot of maize and sorghum), and *F. oxysporum* (*Medicago sativa*, cotton, maize, sweet potato, banana).

It is known that the incidence of *F. moniliforme* increases with continuous cultivation, especially in monoculture maize cropping (Skoglund and Brown 1988). Moisture stress resulting in physiological stress in sorghum increases the incidence of *F. moniliforme* (stalk rot) (Dodman and Obst 1989). This may be because moisture stress reduces plant growth much more severely than fungal growth. For example, *F. oxysporum* f. sp. *melonis* had been observed to attain maximum growth at −2 MPa water potential, and the growth was still evident at −10 MPa (Jorge-Silva et al. 1989), suggesting that the fungus can grow in relatively dry soils that restrict plant growth. Thus, soil physical degradation that accompanies land clearing and cultivation which adversely affects soil water regime (infiltration, transmission, and storage) would lead to an increased incidence of *F. moniliforme* and probably many other *Fusarium* diseases.

Fusarium graminearum Schwabe Group I causes crown rot of cereals; it causes economic yield loss, especially in wheat grown in Vertisols and calcareous Alfisols, Entisols, and Inceptisols in the semi-arid subtropical regions. Similar to F. *oxysporum*, the incidence of *F. graminearum* infection increases rapidly when the crop is subjected to water stress, especially just before anthesis and then after up to the grain-filling stage in wheat (Wildermuth et al. 1984). Application of fertiliser N increased the disease incidence in a calcareous Alfisol, presumably by increasing water stress to the crop, whereas gypsum application reduced the disease incidence, especially the deadheads, from 22% without gypsum to 9% (Wildermuth et al. 1994). This is presumably due to improved water infiltration and hence enhanced soil water regime, which may have been adversely affected by organic matter loss and hence aggregate breakdown (Dalal and Mayer 1986a).

Phytophthora Diseases

Plant diseases caused by *Phytophthora* spp. in the tropical and subtropical regions include root rot of *Medicago sativa* and chickpea (*Phytophthora megasperma* f. sp. *medicaginis*), soybean (*P. megasperma* f. sp. glycinea) and safflower (*P. cryptogea*).

Plant root rots caused by *Phytophthora* spp. are most severe when the oxygen supply becomes depleted due to waterlogged conditions (Heritage and Duniway 1985; Allmaras et al. 1988) and where the crop has been preceded by another susceptible crop (Schmitthener 1982, 1985). Topsoil and sub-soil compaction promotes higher infection by *P. megasperma* f. sp. *glycinea* in soybeans (Fulton et al. 1961), probably through its effect on increasing bulk density (Gray and Pope 1986) and thus restricting water movement and causing waterlogging.

Sclerotium spp.

Stem rot and root rot are caused by *Sclerotium* spp. on rice (S. *oryzae*) and peanut (*S. rolfsii*). The incidence of Sclerotial diseases increases with soil acidification (Das et al. 1987), which is a major process of soil degradation (Sect. 3.4).

Table 4.15 Period of cultivation, number of wheat crops and population of Pratylenchus thornei in a Typic Pellustert at two depths

Period of cultivation (year)	Number of wheat crops	Population of P. thornei number kg^{-1} soil	
		0 – 0.1 m	0.1 – 0.2 m
0	0	0	0
2	2	0	0
5	4	0	0
20	11	1705	2665
35	22	3370	3675
50	35	545	670
55	42	1975	1470

Source: Calculated from J.P. Thompson, unpublished data

4.5.2.2 Root-Parasitic Nematodes

Root-feeding nematodes may be migratory ectoparasites, migratory endoparasites, semi-endoparasites, or sedentary endoparasites (Freckman and Caswell 1985).

Root-parasitic nematode, *Rotylenchulus semiformis*, establishes permanent feeding sites within rhizobium nodules on soybeans (Meredith et al. 1983). *Scutellonema cavenessi* alters the physiology of peanut and soybean in that it alters the establishment and function of the symbiosis of the plant with both mycorrhizae and *Rhizobium* (Germani et al. 1980; Germani et al. 1981).

Thompson et al. (1995) found that *Pratylenchus thornei* decreased N uptake by wheat, probably by restricting root growth; infected crops also extracted less water from the soil. He also found that the population of *P. thornei* increased with the period of cultivation and wheat cropping (Table 4.15). *P. thornei* increased to a high population by 35 years of continuous cultivation where 22 wheat crops were grown in a Waco soil (Typic Pellustert).

It has been demonstrated that SOM declined after the soil was cleared of natural vegetation and continuously cultivated for cereal cropping (Sect. 4.1). Since nematode-trapping fungi appear to be most abundant in soils with rich sources of decaying organic matter (Freckman and Caswell 1985), especially macroorganic matter (particulate organic matter), the loss of SOM and more importantly that of macroorganic matter (fPOM and oPOM) accompanying soil degradation (Sect. 4.1) may contribute towards increased virulence of *P. thornei* besides the growing of susceptible crops.

Moreover, nematophagous nematodes are disadvantaged in degraded soils accompanying soil physical degradation since they require at least >60 μm pore space and to be surrounded by water film for their mobility compared to the root-parasitic nematodes, especially endoparasites that can move within roots and root channels.

Unfortunately, there is a paucity of systematic studies on the dynamics of soil biological degradation, including the population dynamics of individual species of organisms, their interactions with other species, biodiversity, and the role of these aspects of biological degradation on soil productivity, environment, and sustainable plant production.

Fertility-Degraded Soil: Soil Health Restoration and Management-Strategies and Approaches

5

Abstract

Soil health restoration includes physical, chemical, biological, and ecological restoration. Most management strategies and approaches are interactive and mutually inclusive. Management strategies for physical restoration include soil organic matter management, reducing erosion using conservation practices, improve soil structure and aeration, and restore soil microbial communities. Integrated nutrient management, amendments for ameliorating acidity (liming, phytoremediation, organic matter), sodicity (gypsum, S, organic matter, phytoremediation, microbiome-remediation), leaching of excess salts, and appropriate cropping systems are used for chemical restoration. Biological restoration includes enhanced biomass C inputs, conservation and regenerative agriculture, and faunal and microbial habitat restoration.

Key words

Soil organic C · Soil physical restoration · Soil chemical restoration · Soil biological restoration · Soil ecological restoration

Land degradation via soil erosion is a major threat to agriculture and food production system as it is directly linked with human health and also reflects on the '*One Health concept of Soil-Plant-Animal-Human and Planetary Health*' (Gomiero 2016; Steffen et al. 2015; Lal 2020, 2024) (Fig. 5.1). Steffen et al. (2015) proposed the nine planetary framework boundaries/earth system processes (namely, climate change, land system change, stratosphere ozone depletion, atmospheric aerosol pollution, ocean acidification, freshwater use, biosphere integrity), which significantly influence human development. In addition, Kraamwinkel et al. (2021) suggested that soil degradation is a 10th planetary boundary, which significantly

R. C. Dalal, S. Jayaraman, *Soil Fertility Degradation, and Soil Health Restoration and Management*, https://doi.org/10.1007/978-981-96-0405-0_5

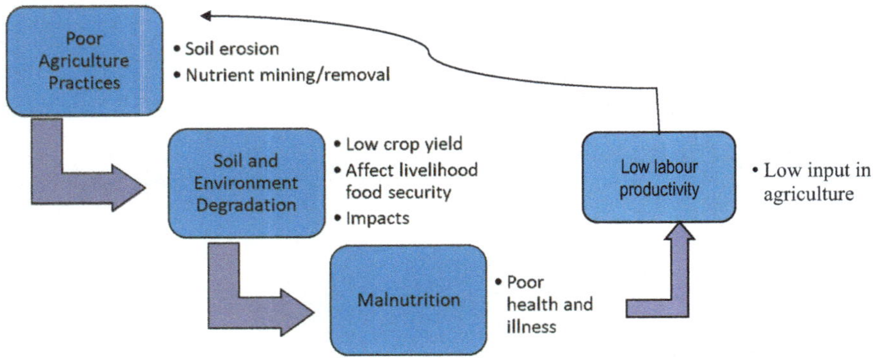

Fig. 5.1 Inter-linking soil degradation and human health. (Modified from Deckelbaum et al. 2006; Lal 2009)

affects food and nutritional security and also the livelihood security of adjoining farming communities. Land degradation via soil erosion (water and wind) is a serious concern as it degrades large areas in many countries especially in the USA (32% area), China (30.7%), Africa (16%), Europe (17%), and Australia (66%). Soil/land degradation occurs through soil physical, chemical, biological, and ecological degradation (Dalal and Bridge 1996; Lal 2009; Lal 2015b), which has a severe impact on food production and sustainability and ecosystem services (Montgomery 2007; Silver et al. 2021; Smith et al. 2021). FAO reported that annually, 75 billion Mg of soil is eroded from arable lands globally, which is equivalent to a monetary loss of US $400 billion annually (FAO, ITPS 2015; FAO 2015, 2017).

Various other kinds of land degradation take place simultaneously. However, dominant processes of fertility degradation (fertility decline or depletion) include soil erosion, nutrient removal in produce, plant residue or biomass removal, oxidation (burning) reduction and/or leaching, acidification, sodification and excessive element load, aggregate breakdown or structural degradation, and loss of microbial biomass and biological diversity (Dalal et al. 1997; Lal 2009, 2021a, b). The nature of soil fertility degradation is the combined effects of one or more processes of fertility degradation, and it is identified via loss of SOM, nutrient depletion, decline in nutrient bioavailability, soil physical degradation, and biological degradation (Figs. 5.1 and 5.2). Similar to land degradation, loss of soil organic matter (SOM) or soil organic carbon (SOC) is a second major threat to farming, soil functions, and ecosystem services (Adhikari and Hartemink 2016; Smith et al. 2021; Silver et al. 2021). Worldwide, about 25–75% SOC of original carbon stock has been lost through unsustainable practices or land use conversion from perennial vegetation to agriculture/arable cropping. It is estimated that 116 billion Mg of SOC has been lost since agriculture began about 12,000 years ago (Sanderman et al. 2018).

In this chapter, restoration and management of fertility-degraded soils are discussed to revert degradation and sustain soil health and food production (Fig. 5.2).

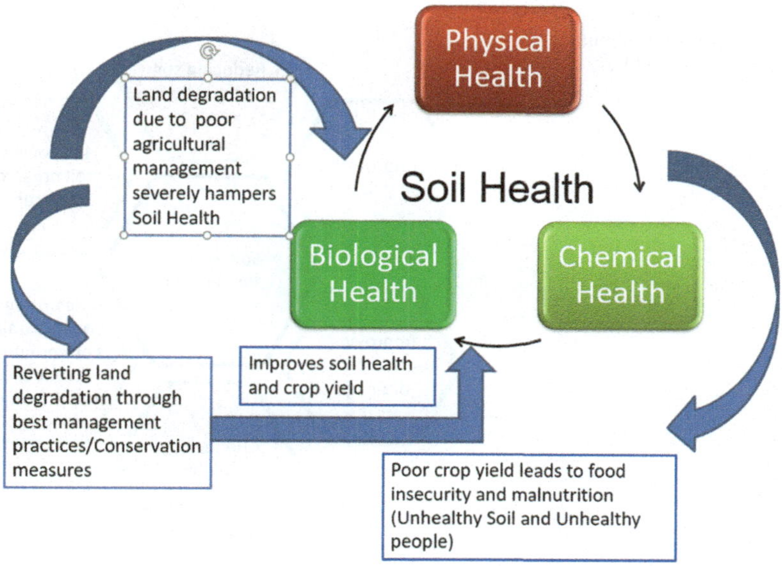

Fig. 5.2 Reversing land degradation through best management practices and improving soil health. (Modified and developed by authors from many literature sources)

5.1 Soil Physical Restoration

In the process of physical degradation of soil, the following issues have been identified, namely, topsoil removal, loss of organic matter, soil (sub-soil) compaction, structural breakdown, poor aeration/moisture retention, poor root penetration, waterlogging due to clogging of pores or poor infiltration, and low hydraulic conductivity (Rasool et al. 2007). Physical degradation mainly occurs due to faulty management practices by non-adoption of appropriate conservation measures or best management practices. Of late, intensive farming practices (intensive tillage, residue burning, no or less organic input addition, absence of crop rotation/cover crop/green manure crop) are one of the major issues which led to structural deterioration and loss of C in the aggregates. Therefore, physical restoration is essential not only for enhancing soil health but also for protecting carbon or storing organic C in soil. The restoration process involves adopting best management practices viz. SOC–centric sustainable soil management practices (SSM) involving conservation agriculture (CA)/no-till (NT) farming (minimizing soil disturbance and residue addition and reducing residue burning and crop rotation including deep-rooted legume crops), regenerative agriculture, erosion control by adopting appropriate site-specific conservation measures (considering soil type, slope, and land use conditions), cropping system (combination of erosion permitting and erosion resistant crops), grazing rotation, and agroforestry/agro-pasture (FAO 2015, 2017; Mosier et al. 2021). One important physical constraint often neglected, i.e. sub-soil constraints, such as

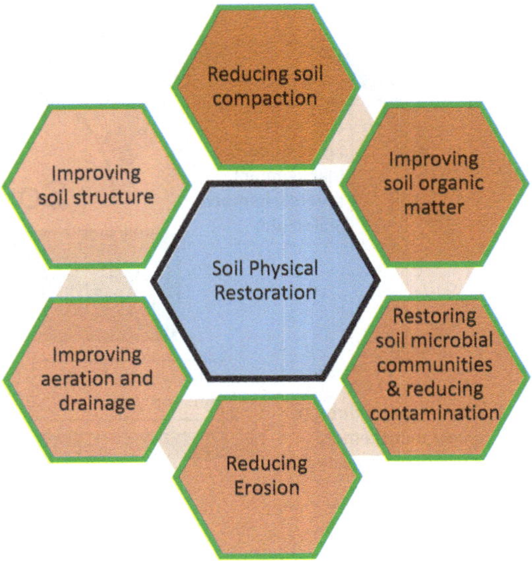

Fig. 5.3 Soil physical restoration through different strategies. (Developed by authors)

improper drainage/sub-soil compaction/accumulation of excess salts, leads to poor crop performance and sometimes may lead to crop failure. Often, removing topsoil constraints by appropriate management practices also reduces the subsoil constraints (Dang et al. 2010).

Strategies commonly used for soil physical restoration are discussed here (Fig. 5.3):

- **Soil organic matter management:** The addition of organic matter through compost, manure, or cover crops can improve soil structure by increasing soil aggregation and porosity, water retention, nutrient availability, and SOC sequestration (FAO 2019). It not only enhances microbial activity and microbial diversity in soil but also helps in improving soil health (Dalal and Bridge 1996; Blanco-Canqui and Lal 2004; Sundermeier et al. 2011; Dalal 2021).
- **Reducing soil erosion through appropriate conservation measures:** Adoption of erosion control measures such as land configuration (contour bunds/graded bunding/ridges and furrow, tied ridges, broad bed and furrow, conservation bench terrace), terracing, contour farming, no-till farming, controlled traffic farming, and planting vegetative buffers can help in reducing runoff and control soil erosion, which is crucial for improving soil physical structure (Lal 2003; Guo et al. 2019; Montgomery 2007; Tullberg et al. 2007; Singh et al. 2020; Blanco and Lal 2023).
- **Reducing soil compaction:** Compaction occurs when soil particles are pressed tightly together, reducing pore space and restricting water infiltration, hydraulic conductivity (HC), and root growth. Therefore, techniques such as reduced tillage, use of cover crops, controlled traffic farming (Tullberg et al. 2007), and avoiding heavy machinery on wet soils can alleviate compaction. Sub-soil

compaction is reduced by appropriate tillage practices (deep tillage to break hardpan) and proper drainage to avoid waterlogging (e.g. mole drains provide efficient drainage in clayey soils under waterlogged conditions) (Spoor 2006; Rao et al. 2009) and also growing of crops having contrasting rooting pattern (soybean + pigeon pea or soybean + cotton). Organic matter amendments after compaction repair slow the reoccurrence of soil compaction.

- **Improving drainage:** Poor drainage can lead to waterlogging and compaction. Installing proper drainage systems or incorporating soil amendments such as gypsum/organic manure can help improve soil drainage (Sharma et al. 2000; Wani et a Rao et al. 2009; Sahrawat et al. 2010; Pathak et al. 2011).
- **Minimizing soil contamination:** Soil contamination can negatively affect soil physical properties and pose risks to human health and the environment. Remediation techniques such as soil washing, bioremediation, and phytoremediation can help mitigate contamination issues (Oliver 1997; Griffiths et al. 2005; Jain et al. 2014; Mishra et al. 2019).
- **Improvement of soil structure:** Practices such as crop rotation, mulching, and avoiding excessive tillage can promote soil aggregation, which enhances soil structure and stability (Six et al. 2000a, b; Sarker et al. 2018; Jayaraman et al. 2017, 2018a, b, c, d, 2021).
- **Soil aeration:** Aerating compacted soils through mechanical means or biological processes can improve soil porosity, root growth, and water infiltration (Kay and Vandenbygaart 2002; Nimmo 2003).
- **Restoring soil microbial communities:** Healthy soil microbial communities play a vital role in maintaining soil structure and nutrient cycling (Jayaraman et al. 2021). Practices like adding microbial inoculants or fostering natural soil microbial populations through organic matter addition can enhance soil restoration efforts (Chander et al. 1997; Harris 2003).
- **Appropriate irrigation management:** Proper irrigation practices, such as drip irrigation or precision irrigation, can help maintain soil moisture levels and prevent waterlogging or drought stress, which can degrade soil physical properties (Rao et al. 2016).

In a nutshell, soil physical restoration often requires a combination of these strategies tailored to specific soil conditions and management goals. In addition, promoting sustainable land management practices and raising awareness about the importance of soil conservation are crucial for ensuring long-term soil health and productivity.

5.2 Soil Chemical Restoration

Soil chemical restoration involves improving the chemical properties of the soil that may have been degraded due to factors such as nutrient depletion, acidification, salinization, sodification, or contamination. Strategies/approaches commonly used for soil chemical restoration are discussed below:

1. **Soil testing and nutrient management:** Conducting soil tests to assess nutrient levels and pH/ electrical conductivity (EC) is essential for developing appropriate nutrient management plans/schemes. Adding fertilizers or soil amendments based on soil test value/results can help replenish depleted nutrients and adjust soil pH to optimal levels for better crop growth (Thakuria et al. 2016; Rashmi et al. 2024).
2. **Organic matter addition:** Incorporating organic matter, such as compost, manure, or crop residues, into the soil can enhance nutrient availability, improve soil structure, and stimulate microbial activity, thereby promoting overall soil health and fertility (Dalal and Bridge 1996; Sarker et al. 2018; Lal 2018; Lorenz and Lal 2018; Henry et al. 2022).
3. **Integrated nutrient management (INM) or integrated plant nutrient system (IPNS):** Replenishing soil nutrients by proper assessment/method is critical for the plant available nutrient sustainability. Therefore, balanced nutrient application through INM or IPNS or through soil test–based recommendation is required (Lakaria et al. 2012; Wu and Ma 2015; Singh et al. 2017, 2021). Adopting 4R techniques (right source, right rate, right time, and right placement) helps in achieving targeted yield goals as well as sustaining soil health. Implementing the 4R approach requires knowledge of soil fertility, crop requirements, and environmental factors, as well as access to tools and technologies for precise nutrient application (Johnston and Bruulsema 2014; Hochmuth et al. 2015). The adoption of integrated soil fertility management (ISFM) is a potential option to reduce the complete dependence on inorganic fertiliser. By following these principles, farmers can optimize fertilizer use, improve crop productivity, and protect natural resources for sustainable agriculture.
4. **Soil Amendments for Ameliorating acidity, Sodicity, or Alkalinity**
   Lime application: Liming acidic soils can raise soil pH, which is important for optimizing nutrient availability and enhancing plant growth and productivity (Rengasamy 2002; Bageria and Baligar 2008; Thakuria et al. 2016; Dhiman et al. 2019; Butterly et al. 2022). Regular liming may be necessary in areas where soil acidification is a problem due to factors like acid rain or intensive agricultural practices.

   Gypsum application: Gypsum application can help improve soil structure and mitigate soil salinity by displacing sodium ions and promoting the flocculation of clay particles, which reduces soil dispersion and enhances water infiltration and salt leaching (Jayaraman et al. 2018a, b, c, d; Rashmi et al. 2024). In areas affected by soil salinization, incorporating amendments such as gypsum, organic matter, or calcium-based fertilizers can help leach excess salts from the soil profile and improve soil structure and fertility (Sharma and Chaudhari 2012; Bageria and Baligar 2008).
5. **Phytoremediation:** Accumulation of excess heavy metal/contaminants always is injurious to plant and soil microbial flora and fauna and subsequent transfer of heavy metals into the food chain to animal and human health (Jayaraman et al. 2005, 2011). Certain plant species, known as hyperaccumulators, have the ability to take up and accumulate in high concentrations of heavy metals or other contaminants from the soil. Planting these species in contaminated areas and

their biomass removal can help reduce soil contamination over time through a process called phytoremediation (Ramana et al. 2009, 2013, 2021). ***Bioremediation***, harnessing the metabolic capabilities of microorganisms to degrade or transform contaminants in the soil, is useful for remediating the contaminated soil. Techniques such as bioaugmentation (adding specialized microorganisms) or biostimulation (providing nutrients or electron acceptors) can enhance the biodegradation of organic pollutants such as pesticides in the soil.

6. **Soil flushing/leaching:** Flushing contaminated soils with clean water or leaching solutions can help remove soluble contaminants and reduce their concentrations in the soil. This technique is particularly effective for soluble salts and certain water-soluble pollutants (Qadir et al. 1998; Nayak et al. 2008; Arnon et al. 2006).
7. **Cover cropping:** Planting cover crops can help improve soil fertility and structure by adding organic matter, reducing erosion, and enhancing nutrient cycling (Lal 2015a). Additionally, certain cover crop species can be used to suppress weeds, nematodes, or soil-borne diseases, reducing the need for chemical inputs.
8. **Site-specific nutrient management through precision farming or sensor-based nutrient application:** Site-specific nutrient management (SSNM) is an approach within precision farming that tailors nutrient application to the specific needs of different areas within a field. This strategy recognizes that soil properties, crop requirements, and environmental conditions can vary significantly across a single field. By utilizing technologies such as GPS, remote sensing, sensors and data analytics, and robotics and artificial intelligence, farmers can precisely assess these variations and apply fertilizers or other nutrients more efficiently (via variable rate application). The benefits of SSNM include increased nutrient efficiency, improved crop yield, and environmental protection (minimizing runoff and nutrient leaching) and also help in achieving sustainability of land use and crop yields.
9. **Long-term monitoring and management:** Regular monitoring of soil chemical properties, nutrient levels, and contaminant concentrations is essential for assessing the effectiveness of soil restoration efforts and making informed management decisions (either reduce the fertiliser dose or increase the recommendation based on target yield goals) (Körschens 2005; Singh et al. 2012; Srinivasarao et al. 2021). Adjustments to management practices may be necessary based on monitoring results and changes in soil conditions over time.

In a nutshell, soil chemical restoration often requires a combination of these strategies tailored to specific soil conditions, management goals, and the type and extent of soil degradation present. Implementing sustainable land management practices and promoting soil conservation are critical for ensuring the long-term health and productivity of soils.

5.3 Soil Biological Restoration

Soil biological restoration focuses on improving the biological components of soil that may have been degraded due to factors such as loss of soil organic matter (SOM), habitat destruction, reduced faunal and microbial biodiversity, contamination/pollution, erosion, or intensive agricultural practices. Commonly used strategies for soil biological restoration include:

1. **Organic matter addition:** Depletion of SOM by inappropriate management or intensive farming practices not only affects soil physical and chemical properties but also biological properties. Therefore, sustaining SOM is critical for achieving targeted yield and also soil health and sustainability. Adding organic materials such as compost, manure, or plant residues to the soil provides food and habitat for soil organisms. This promotes microbial activity, earthworm populations, and other soil fauna, which are essential for nutrient cycling, improvement of soil structure, and overall soil health.
2. **Cover cropping and green manures:** Planting cover crops or green manures can increase biodiversity in the soil by providing a variety of root exudates, organic matter, and habitats for soil organisms (Vukicevich et al. 2016; Muhammad et al. 2021). Different cover crop species have varying effects on soil biology, with options promoting beneficial microbial communities or reducing pest populations.
3. **Crop rotation:** Rotating crops diversifies the root exudates and organic inputs into the soil, which can enhance soil microbial diversity and activity (Jayaraman et al. 2019). Additionally, rotating crops can disrupt pest and disease cycles drastically, reducing the need for chemical inputs and promoting natural biological control mechanisms.
4. **Conservation tillage/conservation agriculture:** Reduced or NT practices help preserve soil structure, organic matter, and soil biota by minimizing disturbance to the soil (Sarker et al. 2018; Hati et al. 2021; Jayaraman and Dalal 2022; Daroch et al. 2024). This allows soil organisms to thrive and perform essential functions such as nutrient cycling, organic matter decomposition, and soil aggregation (Page et al. 2020a, b).
5. **Mulching:** Applying organic or synthetic mulches to the soil surface can provide habitat for soil organisms, regulate soil hydro-thermal regimes, and suppress weed growth. Mulches also contribute to the gradual buildup of organic matter in the soil as they decompose (Lal 2015a, b).
6. **Biochar addition:** Biochar is a carbon-rich material produced from the pyrolysis of organic matter. Adding biochar to the soil can improve soil structure, water retention, and nutrient availability while promoting microbial activity and carbon sequestration in the soil (Hue 2020).
7. **Microbial inoculants:** Introducing beneficial microorganisms such as mycorrhizal fungi, nitrogen-fixing bacteria, or plant growth-promoting rhizobacteria to the soil can enhance plant health, nutrient uptake, and soil fertility (Vessey 2003;

Prasad et al. 2015). Microbial inoculants can be applied as seed treatments, soil amendments, or foliar sprays.

8. **Restoration of soil fauna:** Soil degradation is one of the major causes of biodiversity loss in soil (Kraamwinkel et al. 2021; Weiskopf et al. 2024). Soil fertility degradation, i.e. loss of soil organic carbon/matter and nutrient depletion further exacerbate the situation (Sanchez 2002; Vitousek et al. 2009; Tan et al. 2005). Therefore, reintroducing soil fauna such as earthworms, arthropods, and microarthropods to degraded soils can help restore soil structure, nutrient cycling, and ecosystem functions. Earthworms popularly are known as '*ecosystem engineers*' for their role in the decomposition of organic matter/organic residues and soil aeration. Restoring natural habitats adjacent to agricultural or disturbed areas can provide corridors for soil organisms to migrate and establish populations. Besides, implementing integrated pest management (IPM) practices reduces reliance on chemical pesticides by promoting natural enemies of pests and enhancing biodiversity in agroecosystems. This includes using biological control agents, crop diversification, and habitat manipulation to manage pest populations while minimizing environmental impacts.

Overall, soil biological restoration requires a holistic approach that integrates diverse strategies to enhance soil biodiversity, ecosystem functioning, and resilience to environmental stressors. Prioritizing sustainable land management practices, promoting soil conservation, and restoring SOM in fertility-degraded soils are essential for achieving long-term soil health and productivity.

5.4 Soil Ecological Restoration

Ecological restoration is the practice of repairing and restoring ecosystems that have been degraded, damaged, or destroyed. It envisages to recover the structure, function, and resilience of ecosystems while promoting biodiversity, ecosystem services, and environmental sustainability (Mosier et al. 2021). Ecological restoration is very much essential for achieving better ecosystem services. For achieving these objectives, we need to have proper strategies and approaches such as assessment and planning, understanding native species/conservation of biodiversity, habitat creation (creating or enhancing habitat features such as wetlands, riparian zones, forests, and grasslands can improve ecological connectivity and promote species diversity). In addition, there is a requirement for soil restoration and hydrological restoration, and long-term monitoring and funding (continuous commitment and sufficient funding for planning, implementation, monitoring, and maintenance activities):

1. **Habitat creation and enhancement:** Creating or enhancing habitat features such as wetlands, riparian zones, forests, and grasslands can improve ecological connectivity, promote species diversity, and support ecosystem services such as water filtration, flood control, and carbon sequestration.

2. **Soil restoration:** Improving soil health and fertility through techniques such as soil erosion control, organic matter addition, soil aeration, and soil biota restoration is critical for supporting plant growth, nutrient cycling, and ecosystem functioning/services. **Organic farming and** regenerative agriculture have a role in improving soil health.
3. **Natural regeneration:** Allowing natural processes such as seed dispersal, germination, and succession to occur can facilitate ecosystem recovery in a few cases. Restoring natural disturbances such as fire or flooding, where appropriate, can also promote ecosystem resilience and diversity.
   (i) ***Perennialization*:** It is a strategy aimed at reversing land degradation and promoting sustainable land use by transitioning from annual to perennial vegetation cover (e.g. agro-forestry, sylvi-pasture, and agri-horticulture). It helps in restoring soil fertility, improving C storage/accrual and N and P conservation and availability (Lal 2015b). Moreover, by integrating perennial crops, grasslands, agroforestry systems, and natural vegetation into agricultural landscapes, stakeholders can address soil erosion/degradation, water scarcity, biodiversity loss, and climate change challenges while fostering long-term agricultural productivity and environmental sustainability (Mosier et al. 2021). Incorporating perennial plantation in annual crops brings higher microbial diversity and biomass C in the soil system. Perennialization offers multiple benefits such as improvement in soil health, better ecosystem services via resilience, and agricultural sustainability.
   (ii) **Community participation:** Engaging local communities, stakeholders, and Indigenous people in the restoration process fosters a sense of ownership, builds support for restoration efforts, and integrates traditional ecological knowledge into restoration practices.
4. **Soil Organic Matter Restoration**

Most of the practices briefly stated above involve the optimum restoration of organic matter and soil physical, chemical, and biological functions in fertility-degraded soils (Fig. 5.4). The principles and relative success of these practices in increasing and maintaining SOM and restoring soil health are described below.

5.4.1 Principles of Soil Organic Matter Restoration

Soil organic matter stock comprises 55–58% organic C, 2%–6% N, 0.2%–0.6% P, and variable S concentrations and, on average, with a stoichiometric ratio of 11.5 C:1.0 N:0.15 P:0.15 S (Dalal 1977; Lal 2008; Kirkby et al. 2011; Kopittke et al. 2017; Zhang et al. 2019). It is a function of organic matter C inputs and outputs (Fig. 5.5). The primary source of organic C inputs is the plant matter via photosynthesis, and the outputs are mostly via microbial respiration from organic matter decomposition; minor sources include cyanobacterial crusts, and the outputs are via soil erosion and sediment loss, transport, and leaching. Restoration of SOM or SOC occurs when plant C input, A (Mg plant C ha^{-1} $year^{-1}$) exceeds C output, kC, where

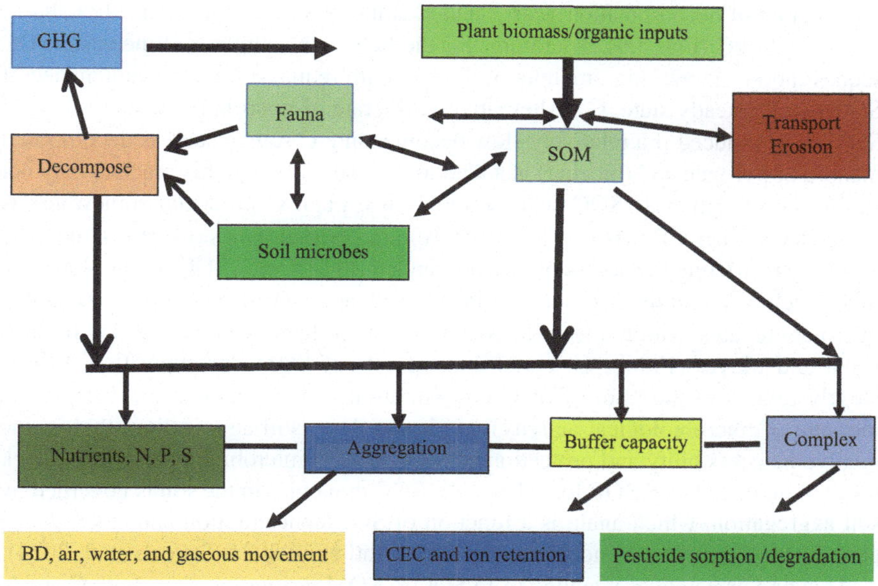

Fig. 5.4 Key functions of soil organic matter (SOM). (Source: authors developed from many literature sources)

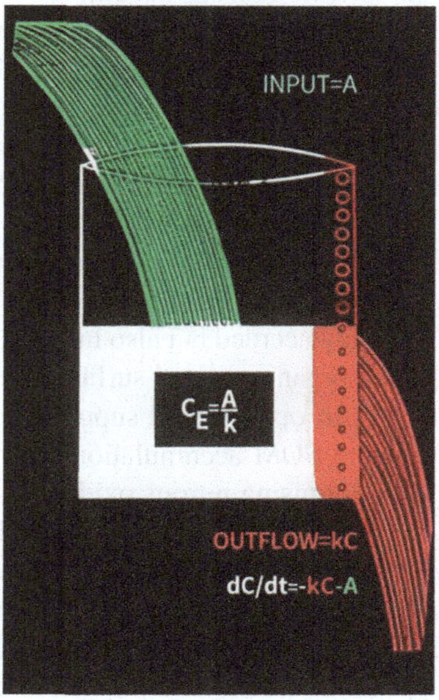

Fig. 5.5 Soil organic C (SOC) stock at any given time is a function of plant C inputs and C outputs, that is, dC/dt and, at steady state, dC/dt = -kC +A or $C_e = A/k$, where C_e is the SOC stock (Mg C ha^{-1}) at steady state, k is the rate of decomposition ($year^{-1}$), A is the rate of plant C inputs (Mg C ha^{-1} $year^{-1}$), and t is the period of continuous management practice in years (year). SOC stock increases when A > kC. (Source: R.C. Dalal unpublished)

k is the rate of decomposition year^{-1} and C is the SOC concentration, Mg C ha^{-1}, that is, A > kC (Fig. 5.9). Therefore, for the SOC concentration to increase (SOC sequestration), increasing amounts of C inputs are required for the maintenance of SOC at new steady state, C_e. Alternatively, the rate of microbial decomposition of SOM, k, is reduced if a relatively slow decomposing C source such as decomposed manure or pyrogenic C (biochar) is added to the soil, although this practice may not strictly be classified as SOC sequestration since plant C produced somewhere is recycled elsewhere (Powlson et al. 2014). Another process that slows down the SOC loss is by restricting the access of microorganisms to SOC and SOC products, that is, SOC occluded within aggregates, oPOM, and aggregated MAOM or occluded micro-aggregates within macro-aggregates (Bhattacharyya et al. 2021). In fact, Dalal and Mayer (1986b) observed that the rate of SOM decomposition, k, was closely related to the ratio of SOC concentration to the soil urease activity (as a measure of microbiological activity) and inversely to soil aggregation; that is, the relative inaccessibility and protection of SOM against microbial and enzyme attack resulted in the reduced SOC loss. Thus, the SOC dynamics in the soil is governed by soil aggregation, which again is a function of SOC concentration and stock (Dalal et al. 2011a, b, c; Somasundaram et al. 2017; Hati et al. 2021; Daroch et al. 2024).

This first principle is extended from SOC to SOM restoration since both C and N in SOM are generally stoichiometric due to the dominant role of microbial decomposers and their products, which form most of the organic matter in the soil (Finn et al. 2015; Kopittke et al. 2020). Consequences of C and N stoichiometry are that microorganisms decomposing organic inputs high in organic C require N either from N fertiliser (or legume N) when available or from SOM, the latter resulting in the decomposition of existing SOM by microbes to mineralise N (N-mining, priming effect) for new microbial-derived SOM. Net SOM increase, especially in MAOM fraction, occurs when the soil contains low total organic C and has a low C:N ratio and relatively high microbial biomass C, while net loss of SOM as MAOM occurs when the soil microbial biomass is relatively high but have low total soil N or increasing C:N ratio of plant C inputs (cereal crop residues or grass litter). That is why N application (either from fertiliser or legume N) not only protects the SOM that is already accrued but also increases the SOM by producing new MAOM by creating N-rich organo-mineral surfaces for further retention of SOM (Kopittke et al. 2018). However, optimised N supply is essential not only for plant production but also to promote SOM accumulation by increasing microbial necromass (Lev et al. 2024) and minimising nitrous oxide emissions from soil (Dalal et al. 2003).

The MAOM fraction provides the long-term storage of SOM (Kleber et al. 2015). However, MAOM is subjected to the C saturation of fine silt+clay (<53 μm) or fine mineral fraction, which is dependent on silt+clay contents and their mineralogy; Fe and Al (hydro-) oxides; specific surface area; soil architecture; nature of organic C inputs, especially their C and N contents; and soil pH. Once the C saturation of mineral fraction is achieved, further SOC (SOM) sequestration for the long-term storage as MAOM is not likely to occur, although the potential turnover through C mineralisation and fresh C addition may still be required (C_e = -kC + A, Fig. 5.9).

Feng et al. (2013) estimated that the silt+clay size fraction (<2 μm) may store 84 ± 1 g C kg^{-1} silt+clay content in 2:1 clay dominant (smectite, illite, vermiculite) soil and 43 ± 1 g C kg^{-1} silt+clay content in 1:1 clay dominant (kaolinite) soil. This provides a 'rule of thumb' estimate to identify a soil, in which SOC may be sequestered long-term in the MAOM fraction. It is worth noting here that the 20–53 μm MAOM fraction may contain silt-size micro-aggregates which may have a faster turnover rate than the <20 μm MAOM. Further C inputs will be stabilized in the oPOM fraction (occluded in aggregates) or remain fPOM. However, oPOM is readily lost when the soil is disturbed, and aggregates break down as it happens when land (soil fertility) degradation processes are dominant in an agroecosystem. It is not known whether there is a saturation limit for the oPOM fraction. However, the fPOM fraction can be increased in the soil indefinitely, although this fraction is affected by climate, global warming, and quantity and quality of continuous C inputs more than the other SOM fractions (Lugato et al. 2021; Rocci et al. 2021) since decomposition rate of fPOM is much faster than the other SOM fractions (fPOM > oPOM > MAOM).

5.4.2 Management Practices for Potential Increase in Soil Organic Matter

Management practices which provide an environment for the formation of macro-aggregates from organo-mineral interactions and micro-aggregation also lead to an increase in SOM (Six et al. 2002; Conrad et al. 2018; Dalal et al. 2018; Page et al. 2020a, b; Hati et al. 2021; Daroch et al. 2024). These include NT and CA, agriculture intensification, intercropping and cover crops, crop-pasture rotations, managed sown pastures, native pastures (rangelands), and crop-pasture-plantation (agroforestry) (Dalal and Somasundaram 2021). Changes in management practice and land use lead to new SOC stock equilibrium (Fig. 5.6).

The contribution of new plant C inputs to SOC is calculated using the ^{13}C natural abundance approach and mixed modelling when $\delta^{13}C$ of the vegetation from previous land use, for example, C_3 vegetation (tree/shrub/forest/legume), differs from the current land use, for example, C_4 vegetation (most pasture grasses, maize, sorghum). This is illustrated for the SOC from native vegetation (Mulga, *Acacia aneura*) to grass pasture (buffel grass, *Cenchrus ciliaris*) or cropping (wheat, *Triticum aestivum*; sorghum, *Sorghum bicolor*) (Fig. 5.7) (Dalal et al. 2021).

5.4.2.1 No-Till Management and Conservation and Climate-Smart Agriculture

Since tillage practice breaks down macro-aggregates, minimum soil disturbance avoids the breakdown of macro-aggregates and preserves macro-aggregates and macro-aggregate-associated SOC (Dalal et al. 1986a, b, c; Dalal et al. 2011a, b, c; Somasundaram et al. 2017, 2020; Hati et al. 2021). For example, in a 50-year NT wheat cropping system on a Vertisol in a semi-arid, subtropical region, the NT practice contained 14.3%, and the CT practice had 9.7% macro-aggregates

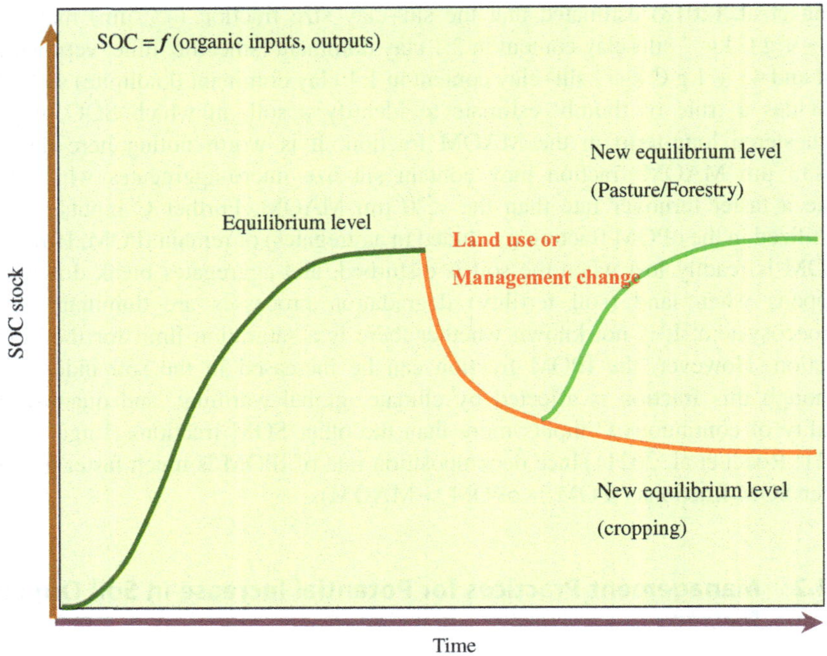

Fig. 5.6 Changes in management practice and land use lead to new SOC stock equilibrium ($C_e = A - kC$, where Ce is the new SOC stock equilibrium, A is the rate of Plant C inputs, and k is the rate of decomposition, see Fig. 5.7 for details). (Source: R.C. Dalal)

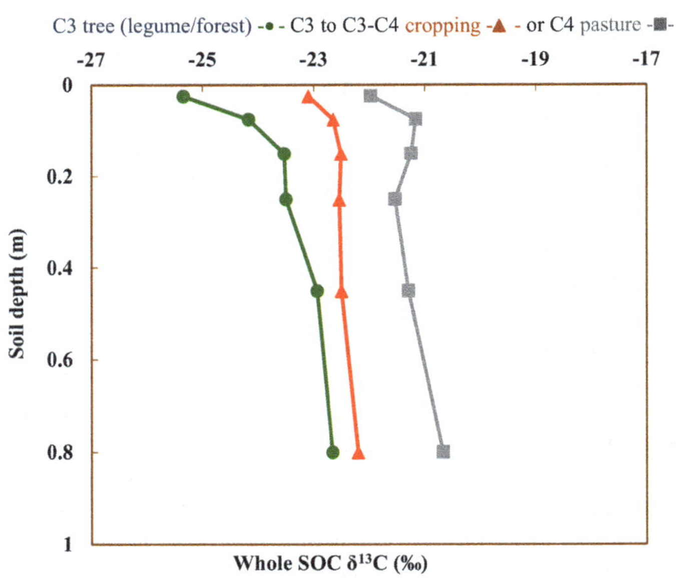

Fig. 5.7 Contribution of plant C inputs to soil organic C (SOC) is apportioned from the plant $\delta^{13}C$ and the SOC $\delta^{13}C$ signatures using mixed models, since C3 plant C $\delta^{13}C$ (−25 to −30‰) differs from C4 plant $\delta^{13}C$ (−12 to −14‰). (Dalal et al. 2005; Richards et al. 2007; Conrad et al. 2018; Dalal et al. 2021)

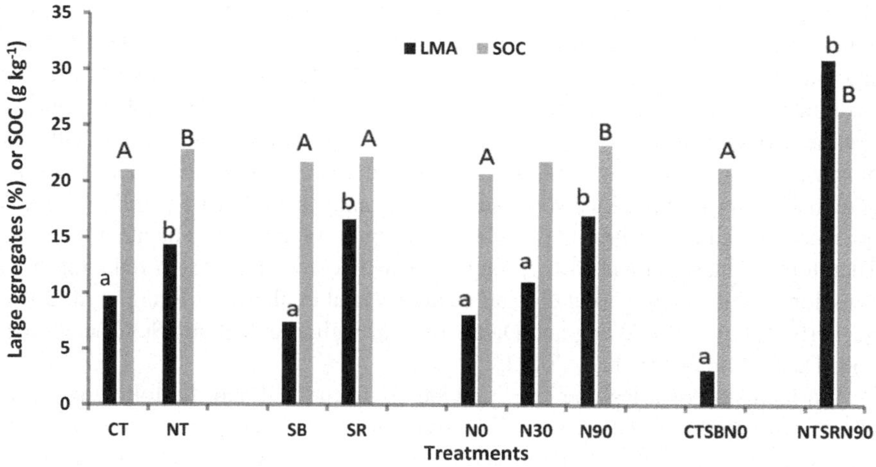

Fig. 5.8 Large aggregates (>2 mm) (black) and associated soil organic C (SOC) concentrations (grey) after 50 years of no-till wheat cropping on a Vertisol (0–0.1 m depth). Treatments are CT = conventional till, NT = no-till, SB = stubble burned, SR = stubble retained, N0 = No fertiliser N applied, N30 = 30 kg N ha^{-1}, N90 = 90 kg N ha^{-1}. Within tillage, stubble, and fertiliser N treatments, different letters (lower case for large aggregates and upper case for SOC) above the bars show a difference at $P < 0.05$. (Adapted and drawn from Hati et al. 2021)

(>2 mm), an increase of 47% from the NT practice, although NT soil had only 10% higher SOC concentration (20.3 g kg^{-1}) than the CT practice (19.3 mg kg^{-1}) (Fig. 5.8). Thus, a small increase in SOC resulted in the formation of much higher proportion of macro-aggregates. Stubble retention led to even much larger differences, 124% higher proportion of large macro-aggregates in the stubble-retained treatment (16.6%) compared to the stubble-burned treatment (9.7%). Furthermore, N fertiliser application (90 kg N ha^{-1}) resulted in 112% higher macro-aggregates (17.0%) compared to the nil N treatment (8.0%). Considering the NT, stubble retention and N treatments as a CA practice, that is, the NT, stubble retained and N fertiliser practice, versus the CT, stubble burned and nil fertiliser practice, the proportion of macro-aggregates in the former was 884% higher and that of macro-aggregate-associated C and N were 24% and 47%, respectively, higher in the former treatment. Therefore, this CA practice not only improved macro-aggregation but also resulted in SOM sequestration (or less SOM loss) in this high clay (68%) soil, in a semi-arid subtropical environment (Hati et al. 2021). Moreover, these practices promoted C and N stoichiometry for the formation and stabilisation of SOM (Finn et al. 2015; Kopittke et al. 2020). Thus, the N-rich microbial necromass creates fresh mineral surfaces for new SOM sorption and the formation of micro-aggregates (Kopittke et al. 2018, 2020) and MAOM, which are occluded within macro-aggregates, since these are protected from accessibility to the microbial decomposition (Dungait et al. 2012; Daroch et al. 2024).

It is worthwhile noting that in soil with low total N, a poor stoichiometry of N to C occurs without additional N supply (fertiliser or legume N), which leads to a priming effect; that is, N mining of old SOM occurs, which leads to the loss of (old) SOC (Chen et al. 2014). Consequently, no net increase in SOM occurs although large amounts of crop residue of wide C:N ratio (>10 Mg ha^{-1}) are added such as green cane trash retention (Page et al. 2013a, b). However, if the accumulation of mineral N either from fertiliser N or legume-derived N occurs in rainfed agriculture with low biomass production in this semi-arid region (Dalal et al. 2018), then there is a high potential for N_2O emissions from soil (Wang and Dalal 2015); thereby, the CA practice may reduce the benefits of SOC sequestration for climate change mitigation. Therefore, N management in cropping systems becomes the key to reducing N_2O emissions, thereby mitigating climate change (Dalal et al. 2003; Wang et al. 2011; Shcherbak et al. 2014; Wang and Dalal 2015) as well as enhancing SOC accumulation (Dalal et al. 2011a, b, c, 2021).

The inclusion of a legume crop provides N to the soil and the following crop, which promotes not only crop productivity but also increases C and N for C sequestration and SOM storage, based on stoichiometry principles (Finn et al. 2015; Kopittke et al. 2020). Besides minimum till or NT and crop residue retention, CA includes crop rotation, with a legume crop that enhances both crop productivity and SOC sequestration by retaining both legume and non-legume crop residues (Hossain et al. 1996; Meena et al. 2018; Kopittke et al. 2020; Somasundaram et al. 2020).

Climate-smart agriculture, a landscape management farming system that integrates cropland, livestock, and fisheries/agro-forestry where appropriate, ensures agricultural productivity and incomes as well as it adapts to climate change while reducing GHG emissions, and this may lead to an increase in SOC sequestration. Even with partial adoption of climate-smart agriculture principles, small but significant increases in SOC stocks compared to the conventional farmers' practice were found in the North-West Indo-Ganga Plains and sub-Saharan Africa (SSA), although there was no significant increase in soil total N stocks, possibly because most SOC increases occurred in the POM fraction (Jat et al. 2019; Kichamu-Wachira et al. 2021). These SOC increases, therefore, could partly be lost by the time of sowing/planting of the next crop. Therefore, optimum N fertiliser application and organic material additions are required in the NT practice to increase SOC/SOM stocks. In spite of the initial low SOC and N stocks in these regions, only a small increase occurred in SOC stocks of <0.5 Mg ha^{-1} in the Indo-Ganga Plains soils and up to 1 Mg ha^{-1} in the SSA soils, that is, on average, 50–100 kg C ha^{-1} year^{-1} in the climate-smart agriculture practised for up to 16 years compared to the CT practice (Powlson et al. 2016). These observations confirm the SOC values obtained from the continuous NT, stubble retention, and N fertiliser application to a Vertisol growing wheat for up to 50 years in the semi-arid subtropical region of Australia (Dalal et al. 2011a, b, c; Somasundaram et al. 2017; Page et al. 2020a, b; Hati et al. 2021); however, the magnitude change of these practices on SOM stocks is climate-dependent. Detailed analysis of CA (including NT) and CT practice showed that a region with a low humidity index (ratio of mean annual precipitation and mean annual temperature) (≤89) (arid to subhumid) is likely to gain SOC (0.40 ± 0.05 Mg ha^{-1} year^{-1}), but a humid region (humidity index, >89) is likely to lose SOC under CA (Sun et al. 2020). Therefore, CA or climate-smart agricultural practices in the long term may enhance SOC sequestration up to 0.4 Mg C ha^{-1} year^{-1} (<8% increase in SOC stocks) (0–0.3 m), although in the semi-arid subtropical and tropical regions, even these increases in the SOM stocks are uncertain (Dalal et al. 2018), and per annum 4 per mille target is less likely to be achieved under current land use practices.

5.4.2.2 Agriculture Intensification, Intercropping, and Cover Crops

Agricultural intensification practices provide additional C inputs to the soil due to the increase in the number and diversity of crops grown per unit of time and, generally, lead to an increase in SOC stocks and improved soil physical, chemical, and biological functions. Provided that N_2O emissions remain low, increasing rates of N application and the inclusion of legumes in an intensified agricultural system using CA practices create a favourable condition for SOC stabilisation and soil aggregation, thus enhancing soil physical, chemical, and biological functions and ecosystem services (Dalal et al. 2003; Shcherbak et al. 2014: Wang and Dalal 2015; Page et al 2020). For example, crop intensification from rice-wheat to rice-wheat-cowpea resulted in the increase in SOC stocks by 0.4 Mg C ha^{-1} year^{-1} over a period of 7 years, but there was no further increase in SOC stocks due to the soil disturbance from potato harvest (Samal et al. 2017). As stated above, however, long-term continuing increases in SOC stocks are uncertain because increasing SOC stocks require increasing C inputs (C_e = A -kC, Fig. 5.9). Moreover, most of the

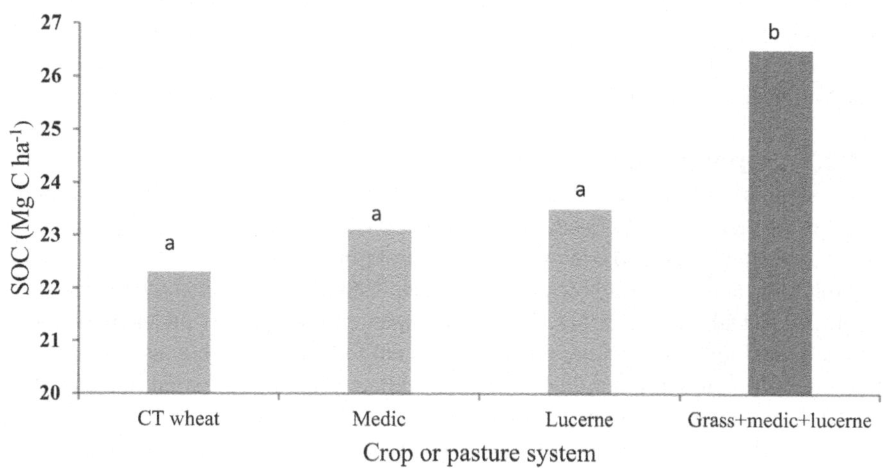

Fig. 5.9 Soil organic C (SOC) stocks four years after land use change from conventional tilled wheat (CT wheat) cropping to annual medic (*Medicago scutellata* and *Medicago truncatula*), Lucerne (*Medicago sativa*), purple pigeon grass (*Setaria incrassate*), and Rhodesgrass (*Chloris gayana*). SOC stocks (0–0.3 m) under grass-legume pasture were 4.2 Mg C ha^{-1} higher than the CT wheat, an average increase of over 1 Mg C ha^{-1} year^{-1}. (Adapted and drawn from Hossain et al. 1996)

short-term increases in SOC stocks are in the oPOM fraction, which is subject to rapid loss following a change in crop and soil management practices.

Growing two complementary crops (maximal/optimal use of space, solar radiation, and nutrient/water acquisition from a shallow and a deep-rooted crop, weed and pest control, soil-borne disease suppressiveness) together maximises the resource use. Examples include the intercropping of pigeon pea (*Cajanus cajan* L.) with maize, which increased SOC concentration as well as total grain protein yield (Dalal 1974; Wang et al. 2014), although the increase in soil total N was generally inconsistent due to N removal in the grain harvest. Even the replacement of rice crops with pigeon peas in the continuous rice-wheat rotations resulted in an increase in SOC concentration. Consequently, wheat yields and C inputs increased due to N_2 fixed by the pigeon pea crop and possibly also to the reduced compaction from puddling used for rice cultivation (Singh and Dwivedi 2006). However, a consistent increase in SOC may not occur in all intercrops (Stomp et al. 2020), especially where biomass production in a sole crop is already high in water-limiting environments. Long-term appropriate intercropping systems are required to increase SOC sequestration when crop residues are retained in situ in the CA system.

Cover crops control weeds, reduce soil erosion, improve soil fertility, control pests and diseases, and promote biodiversity (Lal 2015a). Since cover crops provide biomass additional to the main crop and the residues are retained or used for green and brown manuring, they have the potential to sequester SOC and increase SOM stocks. For example, in long-term experiments, cover crops contributed over 300 kg C ha^{-1} $year^{-1}$ (Bolinder et al. 2020). Also, legume cover crops favour the formation of organo-mineral association and form micro-aggregates, which are then occluded in macroaggregates, resulting in the stabilisation as well as protection of SOC, resulting in increased SOM stocks (Kopittke et al. 2018; Veloso et al. 2019). However, growing of cover crops in rainfed agriculture often experience water supply constraints in semi-arid subtropical and tropical, dryland cropping regions.

5.4.2.3 Crop-Pasture Rotations

In the semi-arid tropical and subtropical regions, where low fertiliser input cropping-livestock systems are practised, grass-legume pastures provide both SOC and total soil N, as an alternative to continuous cereal cropping, which has led to the depletion of both SOC and total soil N (Dalal and Mayer 1986a, b; Dalal et al. 2013; Kopittke et al. 2017; Dalal et al. 2021). Grass accompanied by legumes provides not only N-rich plant biomass for either grazing or animal fodder or both but also plant C and N for the formation of macro-aggregates as well as SOC sequestration (Conrad et al. 2018). Experimentally, it was demonstrated, using ^{13}C- and ^{15}N-labelled residues of Lucerne (*Medicago sativa*) and buffel grass (*Cenchrus ciliaris*), incubated in a Vertisol for a year and analysed using nanoscale secondary ion mass spectrometry (NanoSIMS) that only Lucerne-derived N-rich microbial necromass was directly sorbed on the mineral surfaces, which created new surfaces for further SOC sequestration from microbial decomposed buffel grass C, which was associated with the existing mineral-associated SOC (Kallenbach et al. 2016; Kopittke et al. 2018; Kopittke et al. 2020). Grass-legume pasture systems, therefore, are likely to be more effective in SOC sequestration than either grass or legume-alone pastures (Dalal et al. 1995, 2018; Hossain et al. 1996). Long-term pure grass pasture leads to a decrease in soil total N and productivity (Dalal et al. 2021). Moreover, grass-legume pastures create favourable soil physical and biological conditions due to the formation of macro-aggregates (Conrad et al. 2018).

After long-term cereal cropping, where 60% of the initial SOC was lost, 4 years of pasture phase of growing legumes alone (medic or Lucerne) did not significantly increase SOC stocks. However, in the 4-year grass-legume pasture phase compared to the continuous wheat, SOC stocks increased by 4.2 Mg C ha^{-1} (Fig. 5.9) (Hossain et al. 1996). Continuous wheat crops removed 102 kg N ha^{-1}, whereas grass-legume pasture fixed 384 kg N ha^{-1} during the 4-year period (Hossain et al. 1996; Dalal et al. 2018). Thus, using C and N stoichiometrically of the SOM (ratio of C:N of 11), SOC loss occurred under wheat cropping (1.2 Mg ha^{-1}), but the soil under grass-legume pasture accrued 4.2 Mg C ha^{-1} (Fig. 5.9). Since root biomass contributes to SOM the most, there was a small root biomass under wheat cropping (1.9 Mg dry matter ha^{-1}) compared to the grass-legume pasture (10 Mg dry matter ha^{-1}) in the 0–150 cm profile (Dalal et al. 1995; Dalal et al. 2018). The root biomass of annual medic and Lucerne (<2 Mg ha^{-1}) was similar to the wheat crop (Dalal et al. 1995). Thus, achieving the goal of 4 per mille (0.4% SOC ha^{-1} year^{-1}) for increasing the SOC concentration requires crops/pastures with large root biomass (Chenu et al. 2019). Moreover, increasing crop diversity alone does not necessarily lead to an increase in SOC. For example, King et al. (2020) did not observe increase in SOC after 37 years of maize-soybean rotations; increase in SOC was observed only when a perennial crop/pasture was included in rotation, such as maize-Lucerne, which primarily occurred through increased root C inputs as well as N-rich microbial necromass for SOC stabilisation and SOM storage in macroaggregates (Kopittke et al. 2020).

In a crop-pasture rotation system, the extended grass-legume pasture phase in relation to the cropping phase results in an increase in SOC/SOM stocks. As stated above, crops/pasture with higher root biomass has the potential to increase SOC sequestration (and macro-aggregates) which are critical for sustainable farming systems, although this may be constrained by economic returns and farmer and societal acceptance.

5.4.2.4 Managed Sown Pastures

Managed sown pastures used for livestock grazing include sown grass, legume, or grass-legume pastures. As stated earlier, pure grass pastures are not sustainable in the long term due to N limitation (Dalal et al. 2013). In broad-scale (extensive) pastures, fertiliser use is not sustainable due to low economic return in the semi-arid region of highly variable seasonal and interannual precipitation (Dalal et al. 2013, 2018, 2021). In grass pastures established even on lands of high initial soil N stocks, the pasture productivity gradually declines due to both reduced pasture growth and poor quality (low N concentration) of the pasture grass since plant-available N becomes immobilised in the large grass root biomass (≈10 Mg ha^{-1}). However, in spite of the decrease in total soil N if the grazing intensity is reduced and minimum live or senesced biomass is maintained at >1000 Mg dry matter ha^{-1}, thereby ensuring ground cover and hence reducing SOC loss due to erosion, then the SOC stocks are generally maintained albeit at a much lower level (Dalal et al. 2013; Pringle et al. 2014; Dalal et al. 2021). However, soil N still decreases resulting in poor quality grass. Suitable legumes in the grass pasture are required to provide N via N_2 fixation, which also leads to an increase in SOC sequestration as well as SOM storage.

This principle was proven in Conrad et al.'s (2017) studies, where the Vertisol growing a Leucaena (*Leucaena leucocephala*)-grass (predominantly *Panicum maximum*) pasture for 0, 9, 22, 34, and 40 years increased SOC stocks by 17%–33% in the Leucaena rows compared to grass only (0–0.3 m depth). SOC sequestration rate was 0.28 Mg C ha^{-1} $year^{-1}$ under Leucaena, primarily due to Leucaena fixed N (36 kg N ha^{-1} $year^{-1}$), which promoted the grass productivity, both aboveground and belowground near the Leucaena rows, resulting in the SOC sequestration and SOM storage. The N-rich organic inputs not only create the new organo-mineral surfaces to retain SOC but also increase SOC stabilisation for both non-legume and legume C inputs on the existing organo-mineral surfaces in the soil. In addition, Leucaena stands promoted aggregation, especially the formation of large macro-aggregates (Conrad et al. 2018). SOC stocks were higher in the macro-aggregates than micro-aggregates, because of the higher soil mass distribution rather than a greater C concentration per se. Hence, in Leucaena-grass pastures, macro-aggregates both physically protect SOC per se and act as loci for occluded micro-aggregates to further stabilise SOC and protect it within macro-aggregates (Conrad et al. 2018). Therefore, the introduction/adoption of Leucaena or a legume such as Lucerne or *Desmanthus* spp. in a grass pasture increases pasture productivity, SOC sequestration, and aggregation, hence improving soil physical, chemical, and biological functions in the semi-arid subtropical and tropical regions.

5.4.2.5 Native Pastures

Lands carrying native pastures, also referred to as rangelands or extensive grazing lands, contain mostly native grasses and a sparse population of legumes and shrubs/trees. Rangelands cover > 40% of the Earth's surface and store the largest amount of terrestrial SOC (30%) (Lal 2002; Schuman et al. 2002). Generally, increasing SOC stocks in rangelands is limited due to low biomass production and hence low C inputs due to low and highly variable precipitation/rainfall and high potential evapotranspiration. Moreover, soil N supply (and other nutrients such as P and S) becomes limited, especially if the livestock grazing intensity is high, and nutrients removed in produce and soil erosion are not replaced due to low economic returns constraining the SOC sequestration. Therefore, besides managing animal grazing intensity commensurate with available grazing biomass, N inputs primarily through N_2-fixing pasture legumes and other nutrients such as P and S inputs, stoichiometrically, will maintain or even increase SOC stocks while maintaining productivity (Pringle et al. 2011).

Livestock grazing intensity has the main effect on SOM stocks and SOC sequestration in rangelands (Allen et al. 2013), although greater understanding is required of the belowground processes such as root dynamics and soil nutrient cycling.

Moreover, temperature and rainfall have significant effects on both the amount of C inputs as well as outputs on SOC storage in soils. In addition, utilising the principles of climate-smart agriculture, that is, integration of crop-livestock practices, has demonstrated greater benefits in terms of increasing SOC stocks in the Savanna region of Brazil (de Sant-Anna et al. 2017), although the magnitude of these effects on the SOC stocks varies according to the climate (Abdalla et al. 2018). For example, in the semi-arid subtropical and tropical climates, SOC stocks increased (+7.6%), while in the humid tropical climate, SOC stocks decreased (−19%). Moreover, there was an effect on SOC stock from the species composition in response to grazing intensity. SOC stock increased for C_4-dominated grassland compared to C_3-dominated grassland and C_3–C_4 mixed grasslands at high grazing intensity. Therefore, grazing intensities and management practices should be optimised according to the climate, region, and grassland type to enhance SOC sequestration and increase/maintain the physical, chemical, and biological functions of soil.

Besides grazing intensity, pasture management can have a significant impact on the amounts and long-term sequestration of SOC. Encouragement of pasture growth with high density of roots, root exudation, and low nutrient (N and P) removal leads to SOC sequestration in grazing lands. Practices which lead to the increase in soil N (grass-legume pasture) are closely linked with the increase in SOC stock and SOC sequestration.

It has been recognised that excessive high-intensity grazing leads to the degradation of grassland and SOC loss. For example, in the study using the grazing exclosures in the semi-arid region, grazing exclosures resulted in SOC sequestration of ≈0.04 Mg C ha^{-1} $year^{-1}$ over a 40-year period compared to continuous grazing in an adjacent area, besides an increase in biodiversity in the former (Witt et al. 2011). Admittedly, this practice requires policy shift and societal acceptance for payments for ecosystem services including biodiversity, biomass C and SOC sequestration, and possibly carbon credit for C farming.

5.4.2.6 Crop-Pasture-Plantation-Afforestation

Climate-smart agriculture integrates different land uses (cropping-pasture for livestock-agroforestry or plantation) to optimise resource use. Land use change from perennial vegetation/grasslands to croplands and intensification of agriculture, without or low level of C inputs and nutrients' replacement, deplete SOM stocks (Kopittke et al. 2017; Dalal et al. 2021). Across the range of land uses, grain cropping is the most intensive nutrient and C-extractive activity, while plantation (agroforestry) extracts a small amount of nutrients. Consequently, cropped soil contains the least SOC stock, whereas plantation soil contains the most SOC stock in a similar soil type and climate. Dalal et al. (2021) reported that land use change from native forest to wheat-sorghum cropping decreased SOC stocks by 38% (0.8 Mg C ha^{-1} $year^{-1}$) and to pasture decreased SOC stocks by 12.2% but it decreased soil total N stocks by 24.6% in a Vertisol, due mainly to insufficient C inputs and no N inputs to maintain SOM at steady state (Dalal et al. 2021). In an Alfisol, peanut (*Arachis hypogaea*)-finger millet (*Eleusine coracana*) rotation had an

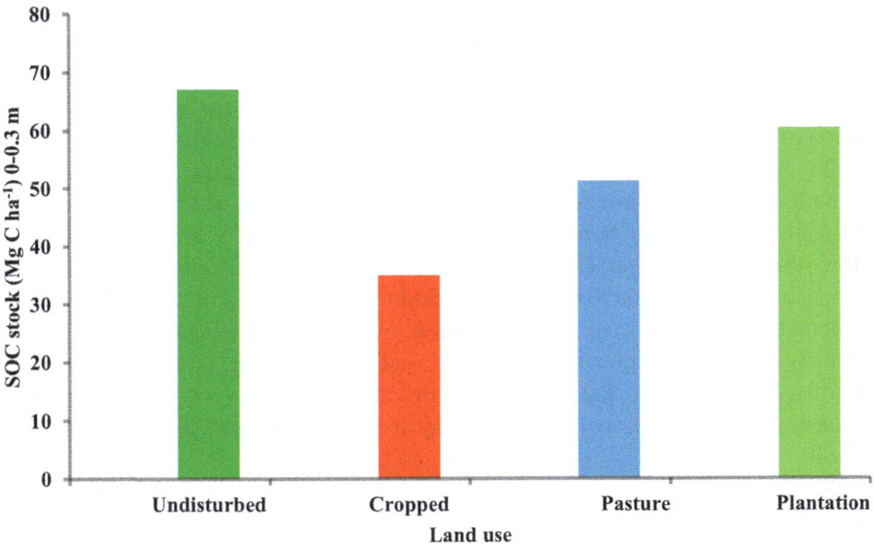

Fig. 5.10 Soil organic C (SOC) stocks (0–1.1 m) of an Oxisol (Tropeptic Eutrustox) under different land uses. Undisturbed native vine scrub land was cropped for 55 years; a part of the cropped land for peanut-maize (after 33y) was converted to pasture (wiregrass (*Aristida ramose*) and Rhodes grass (*Chloris gayana*)) for 22 years; a part of the pastureland (after 18 year) was converted to plantation (*Corymbia maculate*) for 4 years at the time of sampling. (Adapted and redrawn from Mathers et al. 2006)

even higher rate of SOC loss (0.92 Mg C ha^{-1} $year^{-1}$) (Srinivasarao et al. 2014). This is because peanut cropping involves greater intensity of soil disturbance and root C removal and hence leads to faster SOM decomposition due to aggregate disruption and lower C inputs than other crops.

Land use change from a continuously cropped soil to pasture partially restores SOC stocks (Dalal et al. 1995; Hossain et al. 1996; Mathers et al. 2006), and conversion of pastureland to plantation further increases SOC stock, which may become similar to the undisturbed native soil, at least in the top 0.3 m depth (Fig. 5.10). However, SOC stocks in the whole soil profile (0–1.1 m) were only 80% of that in the undisturbed native soil in the semi-arid subtropical Australia (Fig. 5.10) and may require longer period of plantation to restore SOC stock to the original value. Thus, SOC stocks can be restored to the original SOC level in the long-term even in a semi-arid climate following land use change provided that the least disturbed ecosystem is adopted and that it is economically and socially acceptable. These observations are supported by those in similar regions that even a short rotation of Eucalyptus and Leucaena plantations may contribute almost 10 Mg C ha^{-1} $year^{-1}$ to the soil, initially very low in SOC, in the semi-arid tropics of India (Srinivasarao et al. 2014).

Reforestation of agricultural land by afforestation restores the ecosystem and SOC in fertility-degraded lands. Don et al. (2011) observed a significant increase in

SOC accumulation following land use change from cropping to afforestation in the tropics, mimicking the process of bush-fallowing in sub-Saharan Africa. Similarly, Silver et al. (2000) reported a large SOC sequestration potential by reforestation of abandoned tropical agricultural lands. However, the magnitude of SOC stock increase depends on soil type, initial SOC levels, plantation or forest species, and climate. For example, Lewis et al. (2019) found that ecosystem C in the soil reforested for up to 33 years increased by 7.4 Mg ha^{-1} $year^{-1}$ to 0.5 m depth, although SOC increases were inconsistent in the Australian tropics, especially when land use change occurred from long-term (>50 year) established pastures to afforestation or reforestation. Similarly, secondary rainforest land contained a similar amount of SOC stock (138 Mg ha^{-1}) to that in the adjacent pasture (136 Mg ha^{-1}) in the 0–0.3 m depth. Also, the hoop pine plantation of up to 63 year (99.3 Mg C ha^{-1}) was unable to restore SOC stocks to this level possibly due to lower C inputs (Richards et al. 2007) even though SOC sequestration was linked to soil aggregation (Richards et al. 2009).

5.4.2.7 Exogenous Organic C Inputs and Inorganic Amendments (Manure, Compost, Biosolids, Biochar, Microbial Inoculants, Aglime, Gypsum, Silicate Minerals)

Application to the soil of exogenous organic C inputs such as manure is not considered net SOC sequestration since this practice involves the transfer of plant C inputs from one soil (which is disadvantaged due to plant residue removal and subject to fertility degradation) to another soil (which gains plant-derived C consumed by animals or pyrolyzed C) (Powlson et al. 2014). Nevertheless, in the crop-livestock farming system, manure application provides an opportunity to participate in the circular economy, where a substantial portion of plant C and nutrients is recycled through animals on a farm, which otherwise may have been wasted. Application of manure and vermicompost to soil of low SOC and total N stocks in Vertisols and Inceptisols in subtropical India has shown not only an increase in crop productivity but also C sequestration up to 0.18 Mg C ha^{-1} and increased microbial biomass C and N concentrations (Dutta et al. 2024). SOC modelling of long-term experiments (>35 year) on three soils, a Vertisol, an Alfisol, and an Inceptisol in India, showed that farmyard manure (10–15 Mg ha^{-1} $year^{-1}$; ≈2.5–3.5 Mg C ha^{-1} $year^{-1}$) in combination with NPK fertilisers led to 30–107% increase in SOC (Jha et al. 2021). An estimated 20–30% of the farmyard manure C was retained in the soil compared to the control. Retention of added farmyard manure C is much higher than crop residue (0–6% C retained depending on the C:N ratio of the plant N concentration) since a substantial proportion of the C in the decomposed farmyard manure is microbial necromass, hence contributing directly to the SOC.

Biosolids are generally applied to the soil where non-food crops are grown, and plantation and forestry are practised since biosolids may contain pesticides and toxic metals. However, since biosolids are high in nutrients, especially N and P, it is appropriate that these C sources are recycled in a safe environmental practice, as a component of the circular economy.

Application of biochar, a pyrolytic product of organic materials (crop residues, manures, forestry waste, biosolids, and urban waste), may increase SOC, since it contains stable organic C, but also contributes to the soil's cation exchange capacity, especially in coarse-textured soils low in cation exchange capacity, thereby reducing leaching losses of mobile nutrients. Thus, besides an increase in SOC stock, biochar improves other soil conditions such as improving the health of salt-affected soils by increasing aggregate stability, porosity and water retention, cation exchange capacity, SOC and nutrient availability, and microbial biomass as well as alleviating salt stress (Poonia and Parihar 2020; Yuan et al. 2023). Admittedly, biochar is costly for broad-scale agriculture, but it has a niche in specific industries. Bacterial inoculants (such as rhizobium) and fungal inoculants are being used to promote plant growth, enhance plant resistance to various biotic stresses and nutrient cycling, and increase nutrient access to plants (such as AMF) and carbon sequestration via their products and nuclei for soil aggregation (Mendes et al. 2013; Craig et al. 2022; Baron and Rigobelo 2022).

Aglime as calcium (magnesium) carbonate is applied to remediate soil acidity, thereby removing the constraint to crop production, hence leading to an increase in plant C inputs, which may lead to an increase in SOM in the soil. It contributes, however, to the increase in CO_2 fluxes from the decomposition of calcium carbonates and possibly increased soil microbiological activity (Page et al. 2009). Gypsum application with organic manure improves soil aggregation (Niaz et al. 2022) and hence may increase organic matter in soil.

Calcium (magnesium) silicate minerals (such as wollastonite) can be applied to soil, which, on weathering, sequesters CO_2 as an option for climate change mitigation (Haque et al. 2020), although their role in SOM increase is not known. However, it is uncertain as a cost-effective measure due to the mining, transportation and application costs, and slow weathering of the mineral in most soils.

Conclusions

6

Abstract

Soil fertility degradation worldwide is an existential threat to human civilisation. Therefore, holistic approach is required by integrated restoration techniques (soil physical, chemical, and biological) for maintaining a healthy soil -as it is directly linked with one health concept (Soil-Plant-Animal-Human-Planetary health) and helps in achieving Sustainable Development Goals (SDGs).

Key words

Soil fertility degradation · Soil health restoration · Soil stewardship · Sustainable soil management · Integrated restoration techniques · Sustainable development goals

6.1 Soil Stewardship Through Sustainable Soil Management and Restoring Soil Health for Fertility of Degraded Soils

Soil fertility depletion or degradation also referred to as land degradation occurs through loss of fertile topsoil (i.e. removal of topsoil, loss of SOC and plant nutrients, and impaired physical and biological functions) through erosion, crop removal (food grains/fodder), crop residue burning, and inappropriate management practices for cropping, pasture, plantation, and forestry. There is always an imbalance that exists between nutrient removal (crop uptake, harvested through economic produce) and application (replenishment) which eventually leads to fertility depletion in soil. Nutrient replenishment through organic material (crop residue retention/incorporation, green manure, and organic manures) and inorganic fertilisers plays a significant role in maintaining soil fertility in agriculture. Thus, there is an urgent need for sustainable soil management (SSM) practices *involving regenerative agriculture, conservation agriculture (no-till farming/resource conservation*

R. C. Dalal, S. Jayaraman, *Soil Fertility Degradation, and Soil Health Restoration and Management*, https://doi.org/10.1007/978-981-96-0405-0_6

technologies), climate-smart agriculture, organic farming/natural farming (niche areas), and integrated nutrient management (INM)/integrated plant nutrient system (IPNS) module (via 4R techniques/soil test–based nutrient recommendation/site-specific nutrient management via precision farming). In addition, it includes in situ decomposition through microbial consortia to reduce/eliminate residue burning, crop rotation which has contrasting rooting pattern for better nutrient cycling and water use and better carbon storage in the soil profile, soil amendments for ameliorating soil constraints, and microbial consortia for disease suppressiveness. All these strategies/approaches assist in improving soil physical, chemical, and biological fertility, carbon storage, agricultural productivity, and soil health. These strategies also minimise environment degradation (mitigating climate change effects) for better ecosystem function services (supporting, provisioning, regulating, gene pool preserving, and cultural services). Therefore, a holistic approach is needed by integrated restoration techniques (soil physical, chemical, and biological) for maintaining a healthy soil—as it is directly linked with one health concept (*soil-plant-animal-human-planetary health*) and also helps in achieving sustainable development goals (SDGs).

Healthy soil underpins all the 17 SDGs (Smith et al. 2021), as a teacher of one of the authors of this book aptly stated almost six decades ago that *soil is the mother of all resources*. The seventeen SDGs include (United Nations 2015) 1. End poverty in all its forms everywhere (no poverty); 2. End hunger, achieve food security and improved nutrition and promote sustainable agriculture (zero hunger); 3. Ensure healthy lives and promote well-being for all at all ages (good health and well-being); 4. Ensure inclusive and equitable quality education and promote lifelong learning opportunities for all (quality education); 5. Achieve gender equality and empower all women and girls (gender equality); 6. Ensure availability and sustainable management of water and sanitation for all (clean water and sanitation); 7. Ensure access to affordable, reliable, sustainable, and modern energy for all (affordable and clean energy); 8. Promote sustained, inclusive, and sustainable economic growth, full and productive employment, and decent work for all (decent work and economic growth); 9. Build resilient infrastructure, promote inclusive and sustainable industrialisation, and foster innovation (industry, innovation, and infrastructure); 10. Reduce inequality within and among countries (reduce inequalities); 11. Make cities and human settlements inclusive, safe, resilient, and sustainable (sustainable cities and communities); 12. Ensure sustainable consumption and production patterns (responsible consumption and production); 13. Take urgent action to combat climate change and its impacts (climate action); 14. Conserve and sustainably use the oceans, seas, and marine resources for sustainable development (life below water); 15. Protect, restore, and promote sustainable use of terrestrial ecosystems, sustainably manage forests, combat desertification, and halt and reverse land degradation and halt biodiversity loss (life on land); 16. Promote peaceful and inclusive societies for sustainable development, provide access to justice for all, and build effective, accountable, and inclusive institutions at all levels (peace, justice, and strong institutions); 17. Strengthen the means of implementation and revitalise the global partnership for sustainable development (partnerships for the goals).

Healthy soil supports at least twelve SDGs, 1 (no poverty), 2 (zero hunger), 3 (good health and well-being), 5 (gender equality), 6 (clean water and sanitation), 7 (affordable clean energy), 8 (sustainable economic growth), 9 (resilient infrastructure), 11 (sustainable cities), 12 (responsible consumption and production), 13 (climate action), and 15 (life on land) (Keesstra et al. 2018; Lal et al. 2021). This is achieved through restored physical processes (ground cover, good aggregation and porosity and improved aeration, heat and radiation transmission, and water retention, transmission, and filtration), chemical processes (nutrient cycling and retention; remediate acidity, salinity, and sodicity; retain and degrade pesticides and contaminants), biological processes (restore soil organic matter, fauna, microbiome functions, and biodiversity and preserve biological gene pools), and ecosystem services (clean air, clean water, healthy landscapes for human and animal health, aesthetics, cultural values, and generational equity). Most of these goals underpin the practices which promote the restoration of soil organic matter, which is directly responsible for holistic integrated agroecosystem landscapes providing ecosystem function services (supporting, provisioning, regulating, gene pool preserving, and cultural services) to all living beings.

References

ABARE (2009) Australian Bureau of Agricultural and Resource Economics. Australian Farm Surveys – survey methods and definitions 2009

Abdalla M, Hastings A, Chadwick DR, Jones DL, Evans CD, Jones MB, Rees RM, Smith P (2018) Critical review of the impacts of grazing intensity on soil organic carbon storage and other soil quality indicators in extensively managed grasslands. Agric Ecosyst Environ 253:62–81

Adams PB (1990) The potential of mycoparasites for biological control of plant diseases. Annu Rev Phytopathol 28:59–72

Adams TMM, Laughlin RJ (1981) The effects of agronomy on the carbon and nitrogen contained in the soil biomass. J Agric Sci 97:319–327

Adhikari K, Hartemink AE (2016) Linking soils to ecosystem services—a global review. Geoderma 262:101–111

Adimassu Z, Alemu G, Tamene L (2019) Effects of tillage and crop residue management on runoff, soil loss and crop yield in the Humid Highlands of Ethiopia. Agric Syst 168:11–18

Ahlawat IPS (1990) Diagnosis and alleviation of mineral nutrient constraints in chickpea. In: Van Rheenen HA, Saxena MC (eds) Chickpeas in the nineties. International Crops Research Institute for the Semi-arid Tropics, Patancheru, p 93099

Alegre JC (1986) Effect of land clearing and land preparation methods on soil physical and chemical properties aand crop performance of an Ultisol in the Amazon Basin. PhD Thesis Dissertation, North Carolina State University

Alegre JC, Cassell DK (1986) Effect of land-clearing methods and post clearing management on aggregate stability and organic carbon content of a soil in the humid tropics. Soil Sci 142:289–295

Alegre JC, Cassel DK, Bandy DE (1986) Effects of land clearing and subsequent management on soil physical properties. Soil Sci Soc Am J 50(6):1379–1384

Alewell C, Ringeval B, Ballabio C, Robinson DA, Panagos P, Borrelli P (2020) Global phosphorus shortage will be aggravated by soil erosion. Nat Commun 11(1):4546

Al-Kaisi MM (2001) Soil erosion, crop productivity and cultural practices. Iowa State University, University Extension

Allen DE, Pringle MJ, Bray SG, Hall TJ, O'Reagain PO, Phelps D, Cobon DH, Bloesch PM, Dalal RC (2013) What determines soil organic carbon stocks in the grazing lands of north-eastern Australia. Soil Res 51:695–706

Allison FE (1955) The enigma of soil nitrogen balance sheets. Adv Agron 7:213–250

Allison FE (1966) The fate of nitrogen applied to soils. Adv Agron 18:219–258

Allmaras RR, Kraft JM, Miller DE (1988) Effects of soil compaction and incorporated crop residue on root health. Annu Rev Phytopathol 26:219–243

Amézketa E (1999) Soil aggregate stability: a review. J Sustain Agric 14(2-3):83–151

R. C. Dalal, S. Jayaraman, *Soil Fertility Degradation, and Soil Health Restoration and Management*, https://doi.org/10.1007/978-981-96-0405-0

Amoozegar-fard A, Nielsen DR, Warrick AW (1982) Soil solute concentration distributions for spatially varying pore water velocities and apparent diffusion coefficients. Soil Sci Soc Am J 46:3–9

Amundson R, Berhe AA, Hopmans JW, Olson C, Sztein AE, Sparks DL (2015) Soil and human security in the 21st century. Science 348:1261071. https://doi.org/10.1126/s1261071

Anderson DW, Paul EA (1984) Organo-mineral complexes and their study by radiocarbon dating. Soil Sci Soc Am J 48:298–301

Anderson SH, Gantzer CJ, Brown JR (1990) Soil physical properties after 100 years of continuous cultivation. J Soil Water Conserv 45(1):117–121

Andriesse JP (1989) Nutrient management through shifting cultivation. In: van der Heide J (ed) Nutrient management for food crop production in tropical farming systems. Institute for Soil Fertility/Universitas Brawijaya, Haren/Jalan, Malang, pp 29–62

Andriesse JP, Schedhaas RM (1987) A monitoring study of cycles in the soil used for shifting cultivation under various climatic conditions in tropical Asia III. The effects of land clearing through burning on fertility level. Agric Ecosyst Environ 18:311–332

Anton T (2017) Planet of microbes: the perils and potential of earth's essential life forms. University of Chicago Press. https://doi.org/10.7208/chicago/9780226354132.001.0001

Appiah MR, Thompson EJ (1974) The effects of successive cropping on soil organic phosphorus. Ghana J Agric Sci 7:25–30

Armstrong RD, Eagle C, Flood R (2015) Improving grain yields on a sodic soil in a temperate, medium-rainfall cropping environment. Crop Pasture Sci 66:492–505

Arnon S, Ronen Z, Yakirevich A, Adar E (2006) Evaluation of soil flushing potential for clean-up of desert soil contaminated by industrial wastewater. Chemosphere 62(1):17–25

Arriaga FJ, Lowery B (2003) Erosion and productivity. Encyclopedia of Water Science, New York, pp 222–224

Arshad M, Frankenberger WT (1990) Production and stability of ethylene in soil. Biol Fert Soils 10:29–34

Arshad MS, Sanaullah M, Farooq M (2016) Soil carbon sequestration in dryland agriculture. In: Innovations in dryland agriculture, pp 469–490

Atkinson DE (1977) Cellular energy metabolism and its regulation. Academic, New York

Aveyard JM (1983) Soil erosion-productivity research in New South Wales to 1982. Soil Conservation Service, NSW Tech Bull No 24

Ayanaba A, Tuckwell SB, Jenkinson DS (1976) The effects of clearing and cropping on the organic reserves and biomass of tropical forest soils. Soil Biol Biochem 8:519–525

Ayers RS (1977) Quality of water for irrigation. J Irrig Div Proc Am Soc Civil Eng 103:135–154

Azizsoltani E, Honarjoo N, Ayoubi S (2019) How soil pore distribution could help in soil quality studies as an appropriate indicator. Eurasian Soil Sci 52:654–660. https://doi.org/10.1134/S1064229319060036

Bageria NK, Baligar VC (2008) Ameliorating soil acidity of tropical Oxisols by liming for sustainable crop production. Adv Agron 99:345–399

Bageria NK, Baligar VC, Clark RB (2002) Micronutrients in crop production. Adv Agron 77:185–268

Baggs EM (2008) A review of stable isotope techniques for N2O source partitioning in soils: recent progress, remaining challenges, and future considerations. Rapid Commun Mass Spectrom 11:1664–1672

Bai Y, Cotrufo MF (2022) Grassland soil carbon sequestration: current understanding, challenges, and solutions. Science 377(6606):603–608

Baird RW, Knisel WG (1971) Soil Conservation practices and crop production in the black lands of Texas. Conservation Research Reports.15, USDA-ARS, U.S. Gov. Print. Office Washington, DC

Bakker MM, Govers G, Rounsevell MD (2004) The crop productivity–erosion relationship: an analysis based on experimental work. Catena 57(1):55–76

Balasubramanian D, Zhou WJ, Ji HL, Grace J, Bai XL, Song QH, Liu YT, Sha LQ, Fei XH, Zhang X, Zhao JB (2020) Environmental and management controls of soil carbon storage in grasslands of southwestern China. J Environ Manage 254:109810

Balesdent J, Wagner GH, Mariotti A (1988) Soil organic matter turnover in long-term field experiments as revealed by carbon-13 natural abundance. Soil Sci Soc Am J 52:118–124

Balesdent J, Mariotti A, Boisgontier D (1990) Effect of tillage on soil organic carbon mineralization estimated from 13C abundance in maize fields. J Soil Sci 41(4):587–596

Barnwell TO, Jackson RB, Elliott ET, Burke IC, Cole CV, Paustian K, Paul EA, Donigian AS, Patwardhan AS, Rowell A, Weinrich K (1992) An approach to assessment of management impacts on agricultural soil carbon. Natural Sinks of CO2: Palmas Del Mar, Puerto Rico, 24–27 February 1992:423–35

Baron NC, Rigobelo EC (2022) Endophytic fungi: a tool for plant growth promotion and sustainable agriculture. Mycology 13:39–55

Barrow CJ (1991) Land degradation: development and breakdown of terrestrial environments. Cambridge Univ Press, Cambridge

Bartholomew WV, Kirkham D (1960) Mathematical descriptions and interpretations of culture induced soil nitrogen changes. In: Transactions, 7th international congress of soil science communications, vol 111, pp 471–477

Basden, Tom Faulkner J, Sharpley A (2023) How is phosphorus is lost from farmland. Content accessed on 17 April 2023. https://extension.wvu.edu/natural-resources/soil-water/phosphorus-loss

Batey T (2009) Soil compaction and soil management-a review. Soil Use Manag 25:335–345

Batjes NH, Sombroek WG (1997) Possibilities for carbon sequestration in tropical and subtropical soils. Glob Chang Biol 3(2):161–173

Baver LD (1968) The effect of organic matter on soil structure. Pontificia Accademia Scientiarum Scripta Varia 32:383–403

Beegle D, Durst P (2002) Managing phosphorus for crop production, Agronomy facts 13. Penn State Extension

Bengough AG (2003) Root growth and function in relation to soil structure, composition, and strength. In: Root ecology. Springer, Berlin/Heidelberg, pp 151–171

Bengough AG, Young IM (1993) Root elongation of seedling peas through layered soil of different penetration resistances. Plant Soil 149:129–139

Bengough AG, Bransby MF, Hans J, McKenna SJ, Roberts TJ, Valentine TA (2006) Root responses to soil physical conditions: growth dynamics from field to cell. J Exp Bot 57:437–447

Berisso FE, Schjønning P, Keller T, Lamandé M, Etana A, de Jonge LW, Iversen BV, Arvidsson J, Forkman J (2012) Persistent effects of subsoil compaction on pore size distribution and gas transport in a loamy soil. Soil Tilliage Res 122:42–51

Bertol I, Luciano RV, Bertol C, Bagio B (2017) Nutrient and organic carbon losses, enrichment rate, and cost of water erosion. Rev Bras Ciênc Solo 41:e0160150

Bettany JR, Stewart JWB, Saggar S (1979) The nature and forms of sulfur in organic matter fractions of soils selected along an environmental gradient. Soil Sci Soc Am J 43:981–985

Bhattacharyya T, Babu R, Sarkar D, Mandal C, Dhyani BL, Parmar AP (2007) Soil loss and crop productivity model in humid subtropical India. Cur Sci 93:1397–1403

Bhattacharyya R, Rabbi SMF, Zhang Y, Young IM, Jones AR, Dennis PG, Menzies NW, Kopittke PM, Dalal RC (2021) Soil carbon is significantly associated with pore geometry, microbial diversity, and enzyme activity of the macro-aggregates under different land uses of Australia. Sci Total Environ 778:166286

Bhuvaneshwari S, Hettiarachchi H, Meegoda JN (2019) Crop residue burning in India: policy challenges and potential solutions. Int J Environ Res Public Health 16(5):832

Bindraban PS, van der Velde M, Ye L, Van den Berg M, Materechera S, Kiba DI, Tamene L, Ragnarsdóttir KV, Jongschaap R, Hoogmoed M, Hoogmoed W (2012) Assessing the impact of soil degradation on food production. Curr Opin Environ Sustain 4(5):478–488

Blackwell PS, Ringrose-Voase AJ, Jayawardane NS, Olsson KA, McKenzie DC, Mason WK (1990) The use of air-filled porosity and intrinsic permeability to air to characterize structure of macropore space and saturated hydraulic conductivity of clay soils. J Soil Sci 41(2):215–228

Blake L, Mercik S, Koerschens M et al (2000) Phosphorus content in soil, uptake by plants and balance in three European longterm field experiments. Nutr Cycl Agroecosyst 56:263–275. https://doi.org/10.1023/A:1009841603931

Blanco H, Lal R (2023) Soil conservation and management. Springer, Cham

Blanco-Canqui H, Lal R (2004) Mechanisms of carbon sequestration in soil aggregates. Crit Rev Plant Sci 23:481–504

Blanco-Canqui H, Lal R (2009) Crop residue removal impacts on soil productivity and environmental quality. Crit Rev Plant Sci 28:139–163

Blanco-Canqui H, Wienhold BJ, Jin VL, Schmer MR, Kibet LC (2017) Long-term tillage impact on soil hydraulic properties. Soil Tillage Res 170:38–42

Bloem J, Hopkins DW, Benedetti A (eds) (2005) Microbiological methods for assessing soil quality. CABI, pp 1–316

Blonska E, Lasota J, da Silva GR, Vanguelova E, Ashwood F, Tibbett M, Watts K, Lukac M (2020) Soil organic matter stabilization and carbon-cycling enzyme activity are affected by land management. Ann For Res 63(1):71–86

Bloomfleld C, Coulter JK (1974) Genesis and management of acid sulfate soils. Adv Agron 25:265–326

Bohn HL (1976) Estimate of organic carbon in world soils. Soil Sci Soc Am J 40:468–470

Bolin B (1970) The carbon cycle. Scientific American 223:124–132

Bolinder MA, Crotty F, Elsen A, Frac M, Kismányoky T, Lipiec J, Tits M, Tóth Z, Kätterer T (2020) The effect of crop residues, cover crops, manures, and nitrogen fertilization on soil organic carbon changes in agroecosystems: a synthesis of reviews. Mitig Adapt Strateg Glob Chang 25:929–952

Bonde TA, Christensen BT, Cerri CC (1992) Dynamics of soil organic matter as reflected by natural C abundance in particle size fractions of forested and cultivated Oxisols. Soil Biol Biochem 24:275–277

Borrelli P, Robinson DA (2017) An assessment of the global impact of 21st century land use change on soil erosion. Nat Commun 8(1):2013

Borrelli P, Robinson DA, Fleischer LR, Lugato E, Ballabio C, Alewell C, Meusburger K, Modugno S, Schütt B, Ferro V, Bagarello V (2017) An assessment of the global impact of 21st century land use change on soil erosion. Nat Commun 8(1):1–13

Borrelli P, Van Oost K, Meusburger K, Alewell C, Lugato E, Panagos P (2018) A step towards a holistic assessment of soil degradation in Europe: coupling on-site erosion with sediment transfer and carbon fluxes. Environ Res 161:291–298

Bot A, Benites J (2005) The importance of soil organic matter: key to drought-resistant soil and sustained food production (No. 80). Food and Agriculture Organization, Rome, pp 1–78

Bouldin DR, Ritchey KD, Lobato E (1987) Management of soil acidity. In: Sanchez PA, Stoner ER, Pushparajah E (eds) Management of acid tropical soils for sustainable agriculture, Proceedings of an IBSRAM inaugural workshop, Bangkok, Thailand, pp 187–203

Bouwman AF (1989) Modelling soil organic matter decomposition and rainfall erosion in two tropical soils after forest clearing for permanent agriculture. Land Degrad Dev 1(2):125–140

Bouwman AF, Sombroek WG (1990) Inputs to climatic change by soil and agriculture related activities: present status and possible future trends. Dev Soil Sci 20:15–30

Brady NC (1995) The nature and properties of soils. Prentice Hall of India Private Ltd., New Delhi

Brady NC (1996) The nature and properties of soils, 10th edn. Reprinted in India by Special Arrangement with Prentice Hall, Inv, Englewood Cliffs

Bristow KL (1988) The role of mulch and its architecture in modifying soil temperature. Aust J Soil Res 26:269–280

Bromfield SM, Cumming RW, David DJ, Williams CH (1983) Change in soil pH, manganese and aluminium under subterranean clover pasture. Aus J Exp Agric 23(121):181–191

Brookes PC, Newcombe AD, Jenkinson DS (1987) Adenylate energy charge measurements in soil. Soil Biol Biochem 19:211–217

Brownlie WJ, Alexander P, Maslin M, Cañedo-Argüelles M, Sutton MA, Spears BM (2024) Global food security threatened by potassium neglect. Nat Food 19:1–5

Bruce RC, Bell LC, Edwards DG, Warrell LA (1989) Chemical attributes of some Queensland acid soils. II. Relationships between soil and solution phase compositions. Aust J Soil Res 27:353–364

BSES (1979) BSES monitoring sites in the Bundaberg district. Bureau of Sugar Experiment Stations, Brisbane

Buresh RJ, De Datta SK (1990) Denitrification losses from puddled rice soils in the tropics. Biol Fertil Soils 9:1–13

Burford JR, Sahrawat KL (1989) Management of vertisols for improved production. In: Proceeding of the IBSRAM workshop 18–22 Feb, 1985. International Crops Research Institute for the semi-arid tropics, Patancheru

Burford JR, Sahrawat KL, Singh RP (1989) Nutrient management in vertisols in the Indian semi-arid tropics. In: Burford JR, Sahrawat KL (eds) Management of vertisols for improved agricultural production. International Crops Research Institute for the Semi-arid Tropics, Patancheru, pp 147–159

Buringh P (1980) A comparison of three methods for supplying physical data on a crop production for agricultural development. Centre for World Fotui Studies. Res Rep SOW 80-3, Wageningen

Buringh P (1984) Organic carbon in soils of the world. In: Woodwell GM (ed) The role of terrestrial vegetation in the global carbon cycle: measurement by remote sensing, SCOPE. Wiley, Chichester, pp 91–09

Burns RG (1982) Carbon mineralisation by mixed cultures. In: Bull AT, Slater JH (eds) Microbial interactions and communities. Academic, London, pp 415–453

Burns RG, DeForest JL, Marcsen J, Sinsabaugh RL, Stromberger ME, Wallenstein MD, Weintraub MN, Zoppini A (2013) Soil enzymes in a changing environment: current knowledge and future directions. Soil Biol Biochem 58:216–234

Butterly CR, Amado TJ, Tang C (2022) Soil acidity and acidification. In: Oliveira TS, Bell RW (eds) Subsoil constraints for crop production. Springer, Cham, pp 53–81. https://doi.org/10.1007/978-3-031-00317-2_3

Cameron KC, Di HJ, Moir JL (2013) Nitrogen losses from the soil/plant system: a review. Ann Appl Biol 162(2):145–143

Campbell CA (1978) Soil organic carbon, nitrogen, and fertility. In: Schnitzer M, Khan SU (eds) Soil organic matter. Elsevier Scientific Publication, Amsterdam, pp 173–272

Campbell LC (2017) Managing soil fertility decline. In: Nutrient use in crop production. CRC Press, pp 29–52

Campbell CA, Biederbeck VO (1982) Changes in mineral N and numbers of bacteria and actinomycetes during two years under wheat-fallow in southwestern Saskatchewan. Can J Soil Sci 62:125–137

Campbell CA, Zentner RP, Selles F, Biederbeck VO, McConkey BG, Lemke R, Gan YT (2004) Cropping frequency effects on yield of grain, straw, plant N, N balance and annual production of spring wheat in the semiarid prairie. Can J Plant Sci 84:487–501

Campbell CA, Zentner RP, Selles F, Jefferson PG, McConkey BG, Lemke R, Blomert BJ (2005) Long-term effect of cropping system and nitrogen and phosphorus fertilizer on production and nitrogen economy of grain crops in a Brown Chernozem. Can J Plant Sci 85:81–93

Campbell CA, VandenBygaart AJ, Grant B, Zentner RP, McConkey BG, Lemke R, Gregorich EG, Fernandez MR (2007) Quantifying carbon sequestration in a conventionally tilled crop rotation study in southwestern Saskatchewan. Can J Soil Sci 87:23–38

Capriel P, Harter P, Stephenson D (1992) Influence of management on the organic matter of a mineral soil. Soil Sci 153:122–128

Cartwright B, Zarcinas BA, Mayfield AH (1984) Toxic concentrations of boron in a red-brown earth at Gladstone, South Australia. Aust J Soil Res 22:261–272

Cartwright B, Rathjen AJ, Sparrow DHB, Pauli JG, Zarcinas BA (1987) Boron tolerance in Australian varieties of wheat and barley. In: Loughman BC, Gabelman WH (eds) Genetic specificity of mineral nutrition in plants. Martinus Nijhoff Publishers, Dordrecht, pp 139–151

Carver RE, Nelson NO, Roozeboom KL, Kluitenberg GJ, Tomlinson PJ, Kang Q, Abel DS (2022) Cover crop and phosphorus fertilizer management impacts on surface water quality from a no-till corn-soybean rotation. J Environ Manag 301:113818

Cass A, Sumner ME (1982) Soil pore structural stability and irrigation water quality. III. Evaluation of soil stability and crop yield in relation to salinity and sodicity. Soil Sci Soc Am J 46:513–517

Cassman KG, Dobermann A, Walters DT (2002) Agroecosystems, nitrogen-use efficiency, and nitrogen management. AMBIO J Hum Environ 31(2):132–140

Castillo JA, Paez ML (1989) Effect of erosion on productivity of Alfisols. Revista de la Facultad de Agronomia, Universidad Central de Venezuela 37:137–147

Cerri CC, Feller C, Balesdent J, Victoria RL, Plenecassagne A (1985) Application du tracageisotopique natural en C á l'étude de la dynamique de la matiéreorganique dans les sols. CR Acad Sci Paris 300:423–428

Chabalier PF (1985) Fertiliser balances in maize cultivation in the centre of the Ivory Coast. Agron Trop 40:253–260

Chan KY, Bellotti WD, Roberts WP (1988) Changes in surface soil properties of Vertisols under dryland cropping in a semi-arid environment. Aust J Soil Res 26:509–518

Chan KY, Roberts WP, Heenan DP (1992) Organic carbon and associated soil properties of a red earth after 10 years of rotation under different stubble and tillage practices. Soil Res 30(1):71–83

Chander K, Goyal S, Mundra MC, Kapoor KK (1997) Organic matter, microbial biomass, and enzyme activity of soils under different crop rotations in the tropics. Biol Fertil Soils 24:306–310

Charoenchamratcheep C, Smith CJ, Satawathananont S, Patrick WH Jr (1987) Reduction and oxidation of acid sulfate soils of Thailand. Soil Sci Soc Am J 51:630–634

Chatterjee D, Vikramjeet K, Adak T, Kikon ZJ, Mukherjee S, Kuotsu R, Deka BC, Das SR (2022) Modelling release kinetics and path analysis of phosphorus in acid soil as influenced by phosphorus solubilizers. J Soil Sci Plant Nutr 22(4):5049–5067

Chen RM, Senbayram S, Blagodatsky O, Mayachina K, Dittert X, Lin E, Blagodatskaya E, Kuzyakov Y (2014) Soil C and N availability determine the priming effect: microbial N mining and stoichiometric decomposition theories. Glob Chang Biol 20:2356–2367

Chenu C, Angers DA, Barre P, Derrien D, Arrouays D, Balesdent J (2019) Increasing organic stocks in agricultural soils: knowledge gaps and potential innovations. Soil Tillage Res 188:41–52

Chertov OG, Komarov AS (1997) SOMM: a model of soil organic matter dynamics. Ecol Model 94 (2–3):177–189

Chilcott CR, Dalal RC, Parton WJ, Carter JO, King AJ (2007) Long-term trends in fertility of soils under continuous cultivation and cereal cropping in Southern Queensland. IX. Simulation of soil carbon and nitrogen pools using 'CENTURY' model. Aust J Soil Res 45:206–217

Chowdhury SP, Schmid M, Hartmann A, Tripathi AK (2009) Diversity of 16S-rRNA and nifH genes derived from rhizosphere soil and roots of an endemic drought tolerant grass, Lasiurus sindicus. Eur J Soil Biol 45:114–122

Christensen BT (1985) Decomposibility of barley straw: effect of cold-water extraction on dry weight and nutrient content. Soil Biol Biochem 17:93–97

Christensen BT (1992) Physical fractionation of soil and organic matter in primary particle size and density separates. Adv Soil Sci 20:1–90

Clarke GB, Marshall TJ (1947) Influence of cultivation on soil structure and its assessment in soils of variable mechanical composition. J Counc Sci Ind Res Aust 20:162–175

Clarke AL, Russell JS (1977) Crop sequential practices. In: Russell JS, Greacen EL (eds) Soil factors in crop production in a semiarid environment. University of Queensland Press, St Lucia, p 2930295

Cogle AL, Rao KPC, Yule DF, Smith GD, George PJ, Srinivasan ST, Jangawad L (2002) Soil management for Alfisols in the semi-arid tropics: erosion, enrichment ratios and runoff. Soil Use Manage 18:10–17

Cogle AL, Keating MA, Langford PA, Gunton J, Webb IS (2011) Runoff, soil loss, and nutrient transport from cropping systems on Red Ferrosols in tropical northern Australia. Soil Res 49 (1):87–97

Cole GW, Lyles L, Hagen LJ (1983) A simulation model of daily wind erosion soil loss. Trans ASAE 26(6):1758–1765

Coleman K, Jenkinson DS (1996) RothC-26.3—a model for the turnover of carbon in soil. In: Powlson DS, Smith P, Smith JU (eds) Evaluation of soil organic matter models, NATO ASI series, vol 38. Springer, Berlin/Heidelberg. https://doi.org/10.1007/978-3-642-61094-3_17

Coleman DC, Reid CP, Cole CV (1983) Biological strategies of nutrient cycling in soil systems. In: Advances in ecological research, vol 13. Academic, pp 1–55

Coleman K, Jenkinson DS, Crocker GJ, Grace PR, Klir J, Korschens M, Poulton PR, Richter DD (1997) Simulating trends in soil organic carbon in long-term experiments using RothC-26.3. Geoderma 81:29–44

Collins HP, Paul EA, Paustian K, Elliott ET (2019) Characterization of soil organic carbon relative to its stability and turnover. In: Soil organic matter in temperate agroecosystems long term experiments in North America. CRC Press, pp 51–72

Collis-George NC, Lal R (1970) Infiltration into columns of swelling soil as studied by high-speed photography. Aust J Soil Res 8:195–207

Condron LM, Frossard E, Tiessen H, Newman RH, Stewart JWB (1990) Chemical nature of organic phosphorus in cultivated and uncultivated soils under different environmental conditions. J Soil Sci 41:41–50

Conrad KA, Dalal RC, Dalzell SA, Allen DE, Menzies NW (2017) The sequestration and turnover of soil organic carbon in subtropical leucaena-grass pastures. Agric Ecosys Environ 248:38–47

Conrad K, Dalal RC, Fujinuma R, Menzies NW (2018) Soil organic carbon and nitrogen sequestration and turnover in aggregates under subtropical leucaena–grass pastures. Soil Res 56:632–647

Contin M, Corcimaru S, De Nobili M, Brookes PC (2000) Temperature changes and the ATP concentration of the soil microbial biomass. Soil Biol Biochem 32(8–9):1219–1225

Cook GD (1988) Degradation of soil structure under continuous cultivation on the Darling Downs, Queensland. Ph.D. Disser Univ Queensland, Brisbane

Cook AG, Critchley BR, Critchley U, Perfect TJ, Yeadon R (1980) Effects of cultivation and DDT on earthworm activity in a forest soil in the sub-humid tropics. J Appl Ecol 17:21–29

Cook GD, So HB, Dalal RC (1992) Structural degradation of Vertisols under continuous cultivation. Soil Tilliage Res 22:47–64

Cooney DR, Namoi N, Zumpf C, Lim SH, Villamil M, Mitchell R, Lee DK (2023) Biomass production and nutrient removal by perennial energy grasses produced on a wet marginal land. BioEnergy Res 16(2):886–897

Cooper PJ (1972) Aryl sulphatase activity in northern Nigerian soils. Soil Biol Biochem 4(3):333–337

Cotrufo MF, Lavallee JM (2022) Soil organic matter formation, persistence, and functioning: a synthesis of current understanding to inform its conservation and regeneration. Adv Agron. https://doi.org/10.1016/bs.agron.2021.11.002

Coughlan KJ (1984) The structure of Vertisols. In: McGarity JW, Hoult EH, So HB (eds) Properties and utilisation of cracking clay soils, Rev Rural Sci No 5. University of New England, Armidale, pp 87–96

Coughlan KJ, Fox WE, Hughes JD (1973) A study of the mechanisms of aggregation. Aust J Soil Res 11:65–73

Coulombe CE, Wilding LP, Dixon JB (1996) Overview of Vertisols: characteristics and impacts on society. Adv Agron 57:289–375

Coulombe CE, Wilding LP, Dixon JB (2000) Vertisols. In: Sumner ME (ed) Handbook of soil science. CRC Press, New York, pp 269–286

Craig RS, Keller AB, Hobbie SE, Kennedy PG, Weber PK, Pett-Ridge J (2022) Hyphae move matter and microbes to mineral microsites: integrating the hyposphere into conceptual model of soil organic matter stabilization. Glob Chang Biol 28:2527–2540

Craswell ET (1978) Some factors influencing denitrification and nitrogen immobilization in a clay soil. Soil Biol Biochem 10(3):241–245

Craswell ET, Strong WM (1976) Isotopic studies of the nitrogen balance in a cracking clay. III. Nitrogen recovery in plant and soil in relation to the depth of fertilizer addition and rainfall. Soil Res 14(1):75–83

Craswell ET, Waring SA (1972) Effect of grinding on the decomposition of soil organic matter. II. Oxygen uptake and nitrogen mineralization in virgin and cultivated cracking clay soils. Soil Biol Biochem 4:435–442

Critchley BR, Cook AG, Critchley U, Perfect TJ, Russell-Smith A, Yeadon R (1979) Effects of bush clearing and soil cultivation on the invertebrate fauna of a forest soil in the humid tropics. Pedobiologia 19:425–438

Culman S, Zone PP, Kim N et al (2019) Nutrients removed with harvested corn, soybean, and wheat grain in Ohio. Agric Nat Res 2019 Sept 7

Currie JA (1970) Movement of gases in soil respiration. In: Sorption and transport processes in soils, Monograph society of chemistry industry, vol 37, pp 152–171

Curry JP, Good JA (1992) Soil faunal degradation and restoration. Adv Soil Sci 17:171–215

da Silva AP, Kay BD, Perfect E (1994) Factors influencing the least limiting water range of soils. Soil Sci Soc Am J 58:1775–1781

Dalal RC (1974) Effects of intercropping maize with pigeon peas on grain yield and nutrient uptake. Exp Agric 10:219–224

Dalal RC (1975) Urease activity in some Trinidad soils. Soil Biol Biochem 7(1):5–8

Dalal RC (1977) Soil organic phosphorus. Adv Agron 29:83–117

Dalal RC (1979) Mineralization of carbon and phosphorus from carbon-14 and phosphorus-32 labelled plant material added to soil. Soil Sci Soc Am J 43(5):913–916

Dalal RC (1982) Effect of plant growth and addition of plant residues on phosphatase activity in soil. Plant Soil 66:265–269

Dalal RC (1985) Distribution, salinity, kinetic and thermodynamic characteristics of urease activity in a vertisol profile. Soil Res 23(1):49–60

Dalal RC (1989) Long-term effects of no-tillage, crop residue and nitrogen application on properties of a Vertisol. Soil Sci Soc Am J 53:1511–1515

Dalal RC (1990) Soil fertility decline – assessment and management. Proc Int Symp Integr Land Use Manage 11:1–5

Dalal RC (1992) Long-term trends in total nitrogen of a Vertisol subjected to zero-tillage, nitrogen application and stubble retention. Aust J Soil Res 30:223–231

Dalal RC (1998a) On fertility degradation in Vertisols. In: Gopal B, Pathak PS, Saxena KG (eds) Ecology today: an anthology of contemporary ecological research. International Scientific Publications, pp 177–197

Dalal RC (1998b) Soil microbial biomass—what do the numbers really mean? Aust J Exp Agric 38 (7):649–665

Dalal RC (2021) A half century of no-tillage practice for growing cereal crops in subtropical Australia. Agric Sci 32(2):65–69

Dalal RC, Allen DE (2008) Greenhouse gas fluxes from natural ecosystems. Aust J Bot 56(5):369–407

Dalal RC, Bridge BJ (1996) Aggregation and organic matter storage in sub-humid and semi-arid soils. In: Structure and organic matter storage in agricultural soils. CRC Press, pp 263–307

Dalal RC, Carter JO (2019) Soil organic matter dynamics and carbon sequestration in Australian tropical soils. In: Global climate change and tropical ecosystems. CRC Press, pp 283–314

Dalal RC, Chan KY (2001) Soil organic matter in rainfed cropping systems of the Australian cereal belt. Soil Res 39(3):435–464

Dalal RC, Henry RJ (1988) Cultivation effects on carbohydrate contents of soil and soil fractions. Soil Sci Soc Am J 52:1361–1365

Dalal RC, Jayaraman S (2021) Soil organic carbon sequestration and turnover in semi-arid tropics and subtropics in relation to agronomic yield and sustainability. In: Lal R (ed) Soil organic carbon and feeding the future. CRC Press, pp 287–311

Dalal RC, Loch RJ (1984) The effect of erosion processes on nutrient enrichment of sediments eroded from a sandy loam soil. In: National Soils Conference, Brisbane, Australia, p 244

Dalal RC, Mayer RJ (1986a) Long-term trends in fertility of soils under continuous cultivation and cereal cropping in Southern Queensland. I. Overall changes in soil properties and trends in winter cereal yields. Aust J Soil Res 24:265–279

Dalal RC, Mayer RJ (1986b) Long-term trends in fertility of soils under continuous cultivation and cereal cropping in Southern Queensland. II. Total organic carbon and its rate of loss from the soil profile. Aust J Soil Res 24:281–292

Dalal RC, Mayer RJ (1986c) Long-term trends in fertility of soils under continuous cultivation and cereal cropping in Southern Queensland. V. Rate of loss of total nitrogen from the soil profile and changes in carbon-nitrogen ratios. Aust J Soil Res 24:493–504

Dalal RC, Mayer RJ (1986d) Long-term trends in fertility of soils under continuous cultivation and cereal cropping in Southern Queensland. III. Distribution and kinetics of soil organic carbon in particle-size fractions. Aust J Soil Res 24:293–300

Dalal RC, Mayer RJ (1986e) Long-term trends in fertility of soils under continuous cultivation and cereal cropping in Southern Queensland. IV. Loss of organic carbon from different density fractions. Aust J Soil Res 24:301–309

Dalal RC, Mayer RJ (1987a) Long-term trends in fertility of soils under continuous cultivation and cereal cropping in Southern Queensland. VII. Dynamics of nitrogen mineralisation potentials and microbial biomass. Aust J Soil Res 25:461–472

Dalal RC, Mayer RJ (1987b) Long-term trends in fertility of soils under continuous cultivation and cereal cropping in Southern Queensland. VI. Loss of total N from different particle-size and density fractions. Aust J Soil Res 25(1):83–93

Dalal RC, Mayer RJ (1990) Long-term trends infertility of soils under continuous cultivation and cereal cropping in Southern Queensland. VIII. Available nitrogen indices and their relationships to crop yields and nitrogen uptake. Aust J Soil Res 28:563–575

Dalal RC, Probert ME (1994) Nutrient depletion. In: Clarke AL, Wylie PJ (eds) Sustainable agriculture for sub-tropical Australia. Department of Primary Industries, Brisbane, pp 85–105

Dalal RC, Probert ME (1997) Ch. 4. Soil nutrient depletion. In: Clarke AL, Wylie PB (eds) Sustainable Crop Production in the Sub-tropics. QDPI Information Series QI97035. Department of Primary Industries, Brisbane, pp 42–63

Dalal RC, Somasundaram J (2021) Soil organic carbon sequestration and turnover in semi-arid tropics and subtropics in relation to agronomic yield and sustainability. Chapter 13. In: Lal R (ed) Soil organic carbon and feeding the future: basic soil processes, Advanced soil science. CRC Press, Abingdon, pp 288–311

Dalal RC, Strong WM, Weston EJ, Gaffney J (1991) Sustaining multiple production systems. 2. Soil fertility decline and restoration of cropping lands in subtropical Queensland. Trop Grasslands 25:173–180

Dalal RC, Strong WM, Weston EJ, Cooper JE, Lehane KJ, King AJ, Chicken CJ (1995) Sustaining productivity of a Vertisol at Warra, Queensland, with fertilisers, no-tillage, or legumes. 1. Organic matter status. Aus J Exp Agric 35(7): 903–913

Dalal RC, Probert ME, Clarke A, Wylie P (1997) Soil nutrient depletion. In: Clark AL, Wylie PB (eds) Sustainable crop production in the sub-tropics an Australian perspective, pp 42–63

Dalal RC, Strong W, Thomas G, Lehane K, Cooper J, King A (2000) Managing long-term fertility of cropping lands with ley pastures in southern Queensland. Trop Grasslands 34:169–176

Dalal RC, Wang W, Robertson GP, Parton WJ (2003) Nitrous oxide emission from Australian agricultural lands and mitigation options: a review. Aust J Soil Res 41:165–195
Dalal RC, Harms BP, Krull E, Wang WJ (2005) Total soil organic matter and its labile pools following mulga (*Acacia aneura*) clearing for pasture development and cropping 1. Total and labile carbon. Soil Res 43(1):13–20
Dalal RC, Allen DE, Livesley SJ, Richards G (2008) Magnitude and biophysical regulators of methane emission and consumption in the Australian agricultural, forest and submerged landscapes: a review. Plant Soil 309:43–76
Dalal RC, Allen DE, Wang WJ, Reeves S, Gibson I (2011a) Organic carbon and total nitrogen stocks in a Vertisol following 40 years of no-tillage, crop residue retention and nitrogen fertilisation. Soil Tilliage Res 112(2):133–139
Dalal RC, Cowie BA, Allen DE, Yo SA (2011b) Assessing carbon lability of particulate organic matter from δ13C changes following land-use change from C native vegetation to C pasture. Soil Res 49(1):98–103
Dalal RC, Wang W, Allen DE, Reeves S, Menzies NW (2011c) Soil nitrogen and nitrogen-use efficiency under long-term no-till practice. Soil Sci Soc Am J 75(6):2251–2261
Dalal RC, Thornton CM, Cowie BA (2013) Turnover of organic carbon and nitrogen in soil assessed from δ C and δ N changes under pasture and cropping practices and estimates of greenhouse gas emissions. Sci Total Environ 465:26–35
Dalal RC, Strong WM, Weston EJ, Cooper JE, Lehane KJ, King AJ (2018) Farming systems' productivity and soil organic carbon stocks following fertilisers, no-tillage, or legumes on a fertility-depleted soil in a semi-arid subtropical region. Soil Res 5:429–440
Dalal RC, Thornton CM, Allen DE, Owens JS, Kopittke PM (2021) Long-term land use change in Australia from native forest decreases all fractions of soil organic carbon, including resistant organic carbon, for cropping but not sown pasture. Agric Ecosyst Environ 311:107326
Dang YP, Edwards DG, Dalal RC (1992) Prediction of zinc response in wheat from plant tissue and soil tests in Vertisols. In: Fourth national soils conference, vol 110. Australian Society of Soil Science Inc, Adelaide
Dang YP, Dalal RC, Routley R, Schwenke GD, Daniells I (2006) Subsoil constraints to grain production in the cropping soils of the north-eastern region of Australia: an overview. Aus J Exp Agric 46(1):19–35
Dang Y, Dalal RC, Buck SR, Harms BR, Kelly R, Hochman Z, Schwenke GD, Biggs AJW, Ferguson NJ, Norrish S, Routley R, McDonald M, Hall C, Singh DK, Daniells IG, Farquharson R, Manning W, Speirs S, Grewal HS, Cornish P, Bodapati N, Orange D (2010) Diagnosis, extent, impacts, and management of subsoil constraints in the northern grains cropping region of Australia. Aust J Soil Res 48:105–119
Daroch RK, Dang YP, McKenna B, Robinson N, Finn D, Page K, Reeves S, Dalal RC (2024) Biotic relationships to aggregation in a Vertisol under long-term no-till, stubble retention and nitrogen fertilisation. Appl Soil Ecol 195:105260
Daryanto S, Wang L, Jacinthe PA (2017) Meta-analysis of phosphorus loss from no-till soils. J Environ Qual 46(5):1028–1037
Das CM, Mishra SK, Harichandan BK, Narain A (1987) Effect of certain soil types on the growth of *Sclerotium rolfsii* causing stem rot of groundnut. Indian Phytopathol 40:418–419
Das B, Huth N, Probert M, Condron L, Schmidt S (2019) Soil phosphorus modelling for modern agriculture requires balance of science and practicality: a perspective. J Environ Qual 48 (5):1281–1294
Das A, Yadav GS, Layek J, Lal R, Meena RS, Babu S, Ghosh PK (2020) Carbon management in diverse land-use systems of eastern Himalayan subtropics. In: Ghosh P, Mahanta S, Mandal D et al (eds) Carbon management in tropical and sub-tropical terrestrial systems. Springer, Singapore. https://doi.org/10.1007/978-981-13-9628-1_8
Das S, Wang WJ, Reeves S, Dalal RC, Dang YP, Kopittke PM (2022) Combined application of lime and a nitrification inhibitor (3,4-dimethylpyrazole phosphate) markedly decreased nitrous oxide emissions from an acid soil. Agronomy 12:1040

Dash MC, Patra UC (1977) Density, biomass and energy budget of a tropical earthworm population from a grassland site in Orissa, India. Revue d'Ecologie et de Biologie du Sol 14(3):461–471

Davies SL, Newman P, Best B (2010) Impact of soil inversion, soil dilution and claying on non-wetting sandplain soils. Proceedings of 2010 Agribusiness Crop Updates Perth Western Australia (Department of Agriculture and Food, Western Australia and Grains Research and Development Corporation)

Davies S, Boyes T, Reynolds C et al. (2018) Understanding the impact of a comprehensive range of deep tillage practices on crop performance and soil properties for two water repellent soil types. GRDC Grains Research Updates, Perth, Western Australia. www.giwa.org.au/2018researchupdates

De Bruyn LA (1993) Ant composition and activity in naturally-vegetated and farmland environments on contrasting soils at Kellerberrin, Western Australia. Soil Biol Biochem 25 (8):1043–1056

De Bruyn LL, Conacher AJ (1990) The role of termites and ants in soil modification – a review. Soil Res 28(1):55–93

De Leenheer L (1977) Soil fertility study in maximising crop yield on mechanized farms in Belgium: general and agricultural conclusions of a 15-year study (1961–1975). In: Proceedings of the international seminar on Soil Environment and Fertilizer Management. International Agricultural Research Center, Tokyo, pp 133–144

de Sant-Anna SA, Jandali CP, Sá JM, Vilela L, Marcham RL, Alves BJ, Urquiaga S, Boddy RM (2017) Changes in soil organic carbon during 22 years of pastures, cropping or integrated crop/livestock systems in the Brazilian Cerredo. Nutr Cycling Agroecosyst 108:101–120

De Vleeschauwer D, Lal R (1981) Properties of worm casts under secondary tropical forest regrowth. Soil Sci 132(2):175–181

de Vries DA (1975) Heat transfer in soils. In: de Vries DA, Fagan NH (eds) Heat and mass transfer in the biosphere. Wiley, New York, pp 5–28

DeBano LF, Conrad CE (1978) The effect of fire on nutrients in a chaparral ecosystem. Ecology 59 (3):489–497

Deckelbaum RJ, Pam C, Mutua P, DeClerck F (2006) Eco nutrition: implementation models from the Millennium Villages Project in Africa. Food Nutr Bull 27(4):335–342

Deeks LK, Bengough AG, Low D, Billett MF, Zhang X, Crawford JW, Chassell JM, Young IM (2004) Spatial variation of effective porosity and its implications for discharge in an upland headwater catchment in Scotland. J Hydrol 290:217–228

Deng SP, Tabatabai MA (1996) Effect of tillage and residue management on enzyme activities in soils. Biol Fertil Soils 22:208–213. https://doi.org/10.1007/BF00382514

Dent DL, Pons LJ (1995) A world perspective on acid sulphate soils. Geoderma 67(3-4):263–276

Detwiler RP, Hall CAS (1988) Tropical forests and the global carbon cycle. Science 239:42–27

Dexter AR (1987) Mechanics of root growth. Plant Soil 98:303–312

Dexter AR, Czyż EA (2007) Applications of S-theory in the study of soil physical degradation and its consequences. Land Degrad Dev 18(4):369–381

Dhaliwal SS, Naresh RK, Mandal A, Singh R, Dhaliwal MK (2019) Dynamics and transformations of micronutrients in agricultural soils as influenced by organic matter build-up: a review. Environ Sustain Indic 1:100007

Dharmakeerthi RS, Thenabadu MW (1996) Urease activity in soils: a review. J Nat Sci Foundation Sri Lanka 4(3):159–195

Dhillon GS, Van Rees KC (2019) Soil organic carbon sequestration by shelterbelt agroforestry systems in Saskatchewan. Can J of Soil Sci 97(3):394–409

Dhiman D, Sharma R, Samkhya NK, Sepehya S, Sharma SK, Kumar R (2019) Effect of regular application of fertilizers, manure and lime on soil health and productivity of wheat in an acid Alfisol. J Plant Nutr 42(19):2507–2521

Dick RP, Knedler E (2005) Enzymes in soil. In: Hillel D (ed) Encyclopaedia of soils in the environment. Elsevier, Amsterdam, pp 448–456

Dick RP, Tabatabai MA (1986) Hydrolysis of polyphosphates in soils. Soil Sci 142:132–140

Dick WA, Tabatabai MA (1993) Significance and potential uses of soil enzymes. In: Metting FB Jr (ed) Soil microbial ecology: application in agriculture and environmental management. Dekker, New York, pp 95–127

Dick WA, Cost D, Chen L (2008) Availability of sulphur to crops from soil and other sources. In: Sulphur: a missing link between soils, crops, and nutrition, Crop Science Society of America. Soil Science Society of America, Madison, pp 59–82

Dickinson NM (2003) Soil degradation and nutrients. In: The restoration and management of derelict land: modern approaches, pp 50–65

Dinesh R, Chaudhuri SG, Sheeja TE (2006) ATP levels and adenylate energy charge in soils of mangroves in the Andamans. Curr Sci 10:1258–1263

Djindjic F, Börling K, Bergström L (2004) Phosphorus leaching in relation to soil type and soil phosphorus content. J Environ Qual 33(2):678–684

Dodman RL, Obst NR (1989) The role of stress in the development of stalk rot and lodging in grain sorghum. In: Proceedings of the Australian Sorghum Workshop 1989. Toowoomba, Queensland, pp 134–137

Dodor DE, Tabatabai MA (2003) Effect of cropping systems on phosphatases in soils. J Plant Nutr Soil Sci 166(1):7–13

Domsch KH, Beck T, Anderson JP, Soderstrom B, Parkinson D, Trolldenier G (1970) A comparison of methods for soil microbial population and biomass studies, pp 520–533

Don A, Schumacher J, Freibauer A (2011) Impact of tropical land-use change on soil organic carbon stocks – a meta-analysis. Glob Chang Biol 17:1658–1670

Dowling AJ, Yule DF, Lisle AT (1984) The salinisation of Vertisols with shallow water tables in the Emerald irrigation area. In: McGarity JW, Hoult E, So HB (eds) pp 285–291

Dregne HE (1982) Historical perspective of accelerated erosion and effect on world civilization. Am Soc Agron Spec Pub 45:1–14

Dubey RK (2022) Decarbonizing global soils for sustainable development. Antheropol Sci 1:413–416. https://doi.org/10.1007/s44177-022-00033-1

Dungait J, Hopkins D, Gregory A, Whitmore A (2012) Soil organic matter turnover is governed by accessibility not recalcitrance. Glob Chang Biol 18:1781–1796

Dutta D, Meena AL, Bhanu C, Ghasal PC, Choudhary J, Kumar S, Mishra RP, Ansari MA, Raghavendra KJ, Prusty AK, Jat PC, Kashyap P, Punia P, Dixit M, Singh O, Rai AK, Meena AK, Rathi S, Yadav P (2024) Sustainable soil management for climate resilience: long-term management effects on soil carbon sequestration and nitrogen dynamics in a semi-arid tropical Inceptisol of India. J Soil Sci Plant Nutr. https://doi.org/10.1007/s42729-024-01844-4

Duxbury JM, Lauren JG, Fruci JR (1991) Measurement of the biologically active soil nitrogen fraction by a 5N technique. Agric Ecosyst Environ 34:121–129

Dzotsi KA, Jones JW, Adiku SG, Naab JB, Singh U, Porter CH, Gijsman AJ (2010) Modeling soil and plant phosphorus within DSSAT. Ecol Modell 221(23):2839–2849

East G (1985) Erosion survey of cropping areas in the Atherton and Herbert on shires after storm events of 21/11/85 to 26/11/85. Queensland Department of Primary Industries Internal Report, Brisbane

Edwards AP, Bremner JM (1967) Micro aggregates in soils. J Soil Sci 34:105–112

Eiland F (1985) Determination of adenosine triphosphate (ATP) and adenylate energy charge (AEC) in soil and use of adenine nucleotides as measures of soil microbial biomass and activity. Tesseract for PlanteavlsSpecialserie, Copenhagen, pp 1–202

Elliott ET (1986) Aggregate structure and carbon, nitrogen, and phosphorus in native and cultivated soils. Soil Sci Soc Am J 50:627–633

El-Saif SA, Dangler EW (1982) Rainfall erosion in the tropics: a state of the art. In: Soil erosion and conservation in the tropics, ASA Special Publication No. 43. Soil Science Society of America, Madison, pp 1–25

El-Saif SA, Singh S, Pathak P (1987) Physical and conservation constraints and management components for SAT Alfisols. In: Pathak P, El-Saif SA, Singh S (eds) Alfisols in the semi-

arid tropics. International Crops Research Institute for the Semi-arid Tropics, Pancheri, pp 33–48
El-Swaify SA, Dangler EW, Armstrong CL (1982) Soil erosion by water in the tropics. University of Hawaii/HITAHRCTAHR Research and Extension Series 24, Institute of Tropical Agriculture and Human Resources, Honolulu, 173 pp
El-Swaify SA, Arsyad S, Krishna rajah P (1983) Soil erosion by water. In: Carpenter RA (ed) Natural systems for development: what planners need to know. Macmillan Pub Co, New York, pp 99–161
El-Swaify SA, Moldenhauer WC, Lo A (eds) (1985a) Soil erosion and conservation, Proc. of Malama Aina 83. Soil Conservation Society of America, Ankeney. 793 pp
El-Swaify SA, Pathak P, Singh RTJ (1985b) Soil management for optimized productivity under rainfed conditions in the semi-arid tropics. In: Advances in soil science. Springer, New York, pp 1–64
Elwell HA (1981) A soil loss estimation model for southern Africa. In: Morgan RPC (ed) Soil conservation: problems and prospects. Wiley, Chichester, pp 281–292
Esmaeilzadeh-Salesman K, Bahram M, Seraj RG, Gohar D, Omidvar M, Eremeev V, Tager L, Khaleghdoust B, Miracles' SM, Luik A, Loit E (2021) Cropping systems with higher organic carbon promote soil microbial diversity. Agric Ecosyst Environ 319:107521
Eswaran H, van den Berg E, Reich P (1993) Organic carbon in soils of the world. Soil Sci Soc Am J 57:192–194
Evett SR, Agam N, Kustas WP, Colaizzi PD, Schwartz RC (2012) Soil profile method for soil thermal diffusivity, conductivity, and heat flux: comparison to soil heat flux plates. Adv Water Resour 1(50):41–54
Fageria NK, Baligar VC, Clark RB (2002) Micronutrients in crop production. Adv Agron. 77:185–268
Fairbanks BC, Woods LE, Bryant RJ, Elliott ET, Cole CV, Coleman DC (1984) Limitations of ATP estimates of microbial biomass. Soil Biol Biochem 16:549–558
Falloon P, Smith P (2000) Modelling refractory soil organic matter. Biol Fertil Soils 30:388–398. https://doi.org/10.1007/s003740050019
FAO (1979) A provisional methodology for soil degradation assessment. Food and Agriculture Organization, Rome
FAO (1992) FAO production yearbook, vol 46. Food and Agriculture Organization, Rome
FAO (2008) Land and Plant Nutrition Management Service. Rome, Italy
FAO (2015) Revised World Soil Charter, Food and Agriculture Organization of the United Nations Viale dell Terme di Caracalla 00153 Rome
FAO (2017) Voluntary guidelines for sustainable soil management. Food and Agriculture Organization of the United Nations, Rome
FAO (2019) Decarbonization of global soils. In: Food and Agriculture Organization of the United Nations. Intergovernmental Technical Panel on Soils, Rome, pp 1–12
FAO, ITPS (2015) Status of the World's Soil Resources (SWSR) – main report
Fausey NR, Lal R (1990) Soil wetness and anaerobiosis. Adv Soil Sci 11:173–186
Feng W, Plante AF, Six J (2013) Improving estimates of maximal organic carbon stabilization by fine soil particles. Biogeochemistry 112:81–93
Ferraz de Almeida R, Rodrigues Mikhael JE, Oliveira Franco F, Fonseca Santana LM, Wendling B (2019) Measuring the labile and recalcitrant pools of carbon and nitrogen in forested and agricultural soils: a study under tropical conditions. Forests 10(7):544
Field DJ, Minsan B (1999) A description of aggregate liberation and dispersion in a horizons of Australian Vertisols by ultrasonic agitation. Geoderma 91(1–2):11–26
Finck A, Venkatachalam J (1982) Chemical properties and fertility management of Vertisols. In: Symposia Papers II. In: Indian Society of Soil Science (ed) Vertisols and rice soils of the tropics, transactions 12th international congress of soil science, New Delhi, vol 3, pp 61–79

Finn D, Page K, Catton K, Kienzle M, Robertson F, Armstrong R, Dalal RC (2015) Ecological stoichiometry controls the transformation and retention of plant-derived organic matter to humus in response to nitrogen fertilisation. Soil Biol Biochem 99:117–127

Firestone MK (1982) Biological denitrification. In: Nitrogen in agricultural soils, vol 22, pp 289–326

Ford GW, Greenland DJ (1968) The dynamics of partly humified organic matter in some arable soils. Trans 9th Int Congr Soil Sci 2:403–410

Ford GW, Greenland DJ, Odes JM (1969) Separation of the light fraction from soils by ultrasonic dispersion in halogenated hydrocarbons containing a surfactant. J Soil Sci 20:291–296

Foster GR, Young RA, Enabling WH (1985) Sediment composition for nonpoint source pollution analyses. Trans Am Soc Agric Eng 28:133–139

Fowler D, Coyle M, Skiba U, Sutton MA, Cape JN, Reis S, Sheppard LJ, Jenkins A, Grissett B, Galloway JN, Ventouse P (2013) The global nitrogen cycle in the twenty-first century. Philos Trans R Soc B 368(1621):20130164

Frankenberger WT Jr, Dick WA (1983) Relationships between enzyme activities and microbial growth and activity indices in soil. Soil Sci Soc Am J 47:945–951

Freckman DW, Caswell EP (1985) The ecology of nematodes in agroecosystems. Annu Rev Phytopathol 23:275–196

Freebairn DM (2004) Erosion control – some observations on the role of soil conservation structures and conservation. Nat Resour Manage 7:8–13

Freebairn DM, Wicker GH (1986) A study of soil erosion on Vertisols of the eastern Darling Downs, Queensland. I. Effect of surface conditions on soil movement within contour bay catchments. Aust J Soil Res 24:135–138

Freebairn DM, Littleboy M, Smith GD, Coughlan KJ (1991) Optimising soil surface management in response to climatic risk. In: Muchow RC, Bellamy JA (eds) Climatic risks in crop production: models and management for the semi-arid tropics and subtropics. C.A.B. International, Wallingford, pp 283–303

French RJ (1991) Crop monitoring: flowering to harvest. Aust Grain 1:6–7

Freney JR, Simpson JR, Denmead OT, Muirhead WA, Leuning R (1985) Transformations and transfer of nitrogen after irrigating a cracking clay soil with a urea solution. Aust J Agric Res 36:684–694

Fryrear DW, Sutherland PL, Davis G, Hardee G, Dollar M (1999) Wind erosion estimates with RWEQ and WEQ. In: Proceedings of conference sustaining the global farm, 10th International Soil Conservation Organization meeting, Purdue University, pp 760–765

Fuchs M (1986) Heat flux Methods of soil analysis: Part 1, Physical and mineralogical methods, vol 5, pp 957–968

Fulton JM, Mortimore CG, Hildebrand AA (1961) Note on the relation of soil bulk density to the incidence of Phytophthora root and stalk rot of soybeans. Can J Soil Sci 41:247–248

Gabriel JL, Quemada M (2011) Replacing bare fallow with cover crops in a maize cropping system: yield, N uptake and fertiliser fate. Eur J Agron 34(3):133–143

Gantzer CJ, Anderson SH, Thompson AL, Brown JR (1990) Estimating soil erosion after 100 years of cropping on Sanborn field. J Soil Water Conserv 45:641–644

Gardner EA, Coughlan KJ (1982) Physical factors determining soil suitability for irrigated crop production in the Burdekin-Elliot River area, Agric Chem Branch Report No. 20. Department of Primary Industries, Brisbane

Gaydon DS, Wang E, Poulton PL, Ahmad B, Ahmed F, Akhter S, Ali I, Amarasingha R, Chaki AK, Chen C, Choudhury BU et al (2017) Evaluation of the APSIM model in cropping systems of Asia. Field Crop Res 204:52–75

Georgiou K, Jackson RB, Vindušková O, Abramoff RZ, Ahlström A, Feng W, Harden JW, Pellegrini AF, Polley HW, Soong JL, Riley WJ (2022) Global stocks and capacity of mineral-associated soil organic carbon. Nat Commun 13(1):3797

Gerke J (2022) The central role of soil organic matter in soil fertility and carbon storage. Soil Syst 6 (2):33

Germani G, Diem HG, Dommergues YR (1980) Influence of 1, 2 dibromo-3-chloropropane fumigation on nematode population, mycorrhizal infection, N2-fixation and yield of field-grown groundnut. Revue Nématol 3:75–79

Germani G, Ollivier B, Diem HG (1981) Interaction of Scutellonerna cavenessi and Glo~ nus mosseae on growth and N, fixation of soybean. Revue Nématol 4(2):277–280

Ghadiri H, Rose CW (1991) Sorbed chemical transport in overland flow: II. Enrichment ratio variation with erosion processes. J Environ Qual 20:634–641

Gigou J, Dubernard J (1979) Study of the fate of fertiliser nitrogen in a sorghum crop in northern Cameron. In: IAEA (ed) Isotopes and radiation in research on soil-plant relationships. International Atomic Energy Agency, Vienna, pp 121–130

Gigou J, Ganey F, Pichot J (1985) Nitrogen balance in some tropical agro-systems. In: Kang BT, van der Heide J (eds) Nitrogen management in farming systems in humid and subhumid tropics. International Institute of Tropical Agriculture, Ibadan, pp 247–268

Gillman GP (1984) Nutrient availability in acid soils of the tropics following clearing and cultivation. In: Craswell ET, Isbell RF (eds) Proceedings of the international workshop in soils. Australian Centre for International Agricultural Research, Canberra, pp 39–46

Glanville SF, Smith GD (1988) Aggregate breakdown in clay soils under simulated rain and effects on infiltration. Soil Res 26(1):111–120

Goedert WJ (1983) Management of the Cerrado soils of Brazil: a review. J Soil Sci 34(3):405–428

Goedert WJ (1987) Management of acid tropical soils in the savannas of South America. In: Sanchez PA, Stoner ER, Pushparajah E (eds) Management of acid tropical soils for sustainable agriculture: Proceedings of an IBSRAM inaugural workshop, Bangkok, Thailand, pp 109–127

Gokhale NG (1959) Soil nitrogen status under continuous cropping and with manuring in the case of unshaded tea. Soil Sci 87:331–333

Gomiero T (2016) Soil degradation, land scarcity and food security: reviewing a complex challenge. Sustain For 8(3):281

Gourley L (1987) Breeding sorghum for acid soils of the humid tropics. In: Latham M, Ahn P (eds) Land development and management of acid soils in Africa. International Board for soil Research and Management (IBSRM), Bangkok, pp 261–273

Gray LE, Pope RA (1986) Influence of soil compaction on soybean stand, yield and Phytophthora root rot incidence. Agron J 78:189–191

Green PA, Vorosmarty CJ, Meybeck M, Galloway JN, Peterson BJ, Boyer EW (2004) Pre-industrial and contemporary fluxes of nitrogen through rivers: a global assessment based on typology. Biogeochemistry 68:71–105

Greenland DJ (1977) Soil damage by intensive arable cultivation: temporary or permanent? Philos Trans R Soc Lond B 281:193–208

Greenland DJ (1986) Effects of organic matter on the properties of some red soils. In: Li CK et al (eds) Proceedings of the international symposium on red soils. Elsevier Science, Amsterdam, pp 262–273

Greenland DJ, Ford GW (1964) Separation of partially humified organic materials from soils by ultrasonic dispersion. Trans 8th Int Congr Soil Sci 3:137–148

Greenwood DJ (1975) Measurement of soil aeration in soil physical conditions and crop production. Ministry of Agriculture, Fisheries and Food, Technical Bulletin No 29, London, pp 261–275

Griffiths BS, Hallett PD, Kuan HL, Pitkin Y, Aitken MN (2005) Biological and physical resilience of soil amended with heavy metal-contaminated sewage sludge. Eur J Soil Sci 56(2):197–206

Grimme H, Juo ASR (1985) Inorganic nitrogen losses through leaching and denitrification in soils of the humid tropics. Ref Rev 25:57–71

Groenevelt PH, Kay BD, Grant CD (1984) Physical assessment of a soil with respect to rooting potential. Geoderma 34(2):101–114

Groffman PM, Tiedje JM, Robertson GP, Christensen S (1988) Denitrification at different temporal and geographical scales: proximal and distal controls. In: Wilson JR (ed) Advances in nitrogen cycling in agricultural ecosystems. C.A.B. International, Wallingford, pp 174–192

Grundon NJ, Best EK (1982) Survey of the extent of copper deficiency of wheat on the Western Downs, Queensland. Queensland J Agric Anim Sc 39:41–46

Grundon JK, Leslie MJ, Whitehouse EK, Best EK, Wall W, Glasby TJ, Mackenzie J (1985) Nutrient screening of Queensland cereal soils. Queensland Department of Primary Industries, Brisbane

GSP (2017) Global soil partnership endorses guidelines on sustainable soil management. http://www.fao.org/global-soilpartnership/resources/highlights/detail/en/c/416516/. Accessed 14 May 2020

Guerra CA, Heintz-Buschart A, Sikorski J, Chatzinotas A, GuerreroRamírez N, Cesarz S, Beaumelle L, Rillig MC, Maestre FT, Delgado-Baquerizo M, Buscot F (2020) Blind spots in global soil biodiversity and ecosystem function research. Nat Commun 11(1):1–13. https://doi.org/10.1038/s41467-020-17688-2

Guillet B (1982) Study of the turnover of soil organic matter using radioisotopes (C). In: Bonneau M, Souchier B (eds) Constituents and properties of soils. Acad Press, London, pp 238–260

Guo LB, Gifford RM (2002) Soil carbon stocks and land use change: a meta-analysis. Glob Chang Biol 8:345–360

Guo M, Zhang T, Li Z, Xu G (2019) Investigation of runoff and sediment yields under different crop and tillage conditions by field artificial rainfall experiments. Water 11(5):1019

Guo LJ, Zhang ZS, Wang DD, Li CF, Cao CG (2015) Effects of short-term conservation management practices on soil organic carbon fractions and microbial community composition under a rice-wheat rotation system. Biol Fert Soils 51: 65–75

Gupta RK, Abrol IP (1990) Salt-affected soils: their reclamation and management for crop production. Adv Soil Sci 11:223–288

Gupta SR, Rajvanshi R, Singh JS (1981) The role of the termite Odontotermes gurdaspurensis (Isoptera: Termitidae) in plant decomposition in a tropical grassland. Pedobiologia 22:254–261

Haas HJ, Evans CE, Miles EF (1957) Nitrogen and carbon changes in Great Plains soils as influenced by cropping and soil treatments. U.S. Dep Agric Tech Bull No 1164

Habte M, El-Swaify SA (1988) Survival of Rhizobium in an oxisol subjected to incremental simulated erosion. Soil Sci Soc Am J 52(5):1313–1316

Haddad KS, Weir RG (1985) Influence of soil properties on the use of soil and plant zinc to predict zinc response in maize. Aust J Exp Agric 25:856–862

Hagen LJ, Lyles L (1985) Amount and nutrient content of particles produced by aggregate abrasion. In: American Society of Agricultural Engineers (ed) Erosion and soil productivity, ASAE Publication 8-85. American Society of Agricultural Engineers Joseph, Michigan, pp 117–129

Hallsworth EG, Gibbons FR, Lemerle TH (1954) The nutrient status and cultivation practices of soils of the North-West wheat belt of New South Wales. Aust J Agric Res 5:422–447

Haque F, Santos RM, Chiang YW (2020) Optimizing inorganic carbon sequestration and crop yield with wollastonite soil amendment in a microplot study. Front Plant Sci 11:1012

Harris JA (2003) Measurements of the soil microbial community for estimating the success of restoration. Eur J Soil Sci 54(4):801–808

Harte AJ (1984) Effect of tillage on the stability of three red soils of the northern wheat belt. J Soil Conserv Serv NSW 40(2):94–01

Hartmann M, Frey B, Mayer J, Mäder P, Widmer F (2015) Distinct soil microbial diversity under long-term organic and conventional farming. ISME J 9(5):1177–1194

Hati KM, Jha P, Dalal RC, Jayaraman S, Dang YP, Kopittke PM, Kirchhof G, Menzies NM (2021) 50 years of continuous notillage, stubble retention and nitrogen fertilization enhanced macro-aggregate formation and stabilisation in a Vertisol. Soil Tillage Res 214:105163. https://doi.org/10.1016/j.still.2021.105163

Hauck RD (1984) Atmospheric nitrogen chemistry, nitrification, denitrification, and their relationships. In: Hutzinger O (ed) The handbook of environmental chemistry. Springer, Berlin, pp 105–125

Hauser S, Norgrove L (2013) Slash-and-burn agriculture, effects of. Encyclopaedia of Biodiversity 6(2):551–562

Havlin JL (2020) Soil: fertility and nutrient management. In: Landscape and land capacity. CRC Press, pp 251–265

Hayaishi O (2013) Oxygenases, Encyclopedia of biological chemistry. Elsevier, pp 371–374

Haynes RJ (1982) Effects of liming on phosphate and availability in acid soils. A critical review. Plant Soil 68:289–308

Haynes RJ (1984) Lime and phosphate in the soil–plant system. Adv Agron 37:249–315

Haynes RJ, Williams PH (1993) Nutrient cycling and soil fertility in the grazed pasture ecosystem. Adv Agron 49:119–199

He ZL, Yang XE, Stoffella PJ (2005) Trace elements in agroecosystems and impacts on the environment. J Trace Elem Med Biol 19(2–3):125–140

Hebbar P, Berge O, Heulin T, Singh SP (1991) Bacterial antagonists of sunflower (Helianthus annuus L) fungal pathogens. Plant Soil 133:131–140

Henry B, Dalal RC, Harrison MT, Keating B (2022) Creating frameworks to foster soil carbon sequestration (2022) Burleigh Dodds Series in Agricultural Science. Burleigh Scientific Publication, pp 1–43

Henry B, Dalal R, Harrison MT, Keating B (2023) Creating frameworks to foster soil carbon sequestration. In: Rumpel C (ed) Understanding and fostering soil carbon sequestration. Burleigh Dodds Science Publishing Limited, pp 767–808

Heritage AD, Duniway JM (1985) Influence of depleted oxygen supply on phytophthora root rot of safflower in nutrient solution. In: Parker CA, Rovira AD, Moore KJ, Wong PTW (eds) Ecology and management of soil borne plant pathogens. The American Phytopathological Society St Paul, Minnesota, pp 199–202

Herridge DF, Bergersen FJ (1988) Symbiotic nitrogen fixation. In: Advances in nitrogen cycling in agricultural ecosystems, pp 46–65. Ref. 8845–65

Heuer B, Meiri A, Shalevet J (1986) Salt tolerance of eggplant. Plant Soil 95:9–13

Hicks RJ, Stotzky G, van Voris P (1990) Review and evaluation of the effects of xenobiotic chemicals on microorganisms in soil. Adv Appl Microbiol 35:195–253

Hillel D (2007) Soil in the environment: crucible of terrestrial life. Elsevier

Hillel D, Braimoh AK, Vlek PL (2008) Soil degradation under irrigation. In: Land Use and Soil Resources, pp 101–119

Hobbs JA, Thompson CA (1971) Effect of cultivation on the nitrogen and organic carbon contents of a Kansas Argiustoll (Chernozem). Agron J 63:66–68

Hochmuth G, Mylavarapu R, Hanlon E (2015) The four Rs of fertilizer management. Nutrient management of vegetable and row crops handbook. In: Institute of Food and Agricultural Sciences (IFAS). University of Florida, Gainesville, pp 5–7

Hodgson JF (1963) Chemistry of the micronutrient elements in soils. Adv Agron 15:119–159

Hodgson AS, MacLeod DA (1989) Oxygen flux, air-filled porosity, and bulk density as indices of vertisol structure. Soil Sci Soc Am J 53:540–543

Hoffman AS, van Diepen LT, Albeke SE, Williams DG (2022) Potential microbial enzyme activity in seasonal snowpack is high and reveals P limitation. Ecosphere 13(3):e397

Holt JA (1987) Carbon mineralization in semi-arid northeastern Australia: the role of termites. J Tropical Ecol 3(3):255–263

Holt AJ, Lepage M (2000) Termites and soil properties. In: Abe T, Bignell DE, Higashi M (eds) Termites: evolution, sociality, symbioses, ecology. Kluwer Academic Publishers, Dordrecht, pp 389–407

Horn R (1990) Aggregate characterization as compared to soil bulk properties. Soil Tillage Res 17 (3–4):265–289

Hornby D (1983) Suppressive soils. Ann Rev Phytopathol 21:65–85

Hossain SA, Dalal RC, Waring SA, Strong WM, Weston EJ (1996) Comparison of legume-based cropping systems at Warra, Queensland. I. Soil nitrogen and organic carbon accretion and potentially mineralisable nitrogen. Aust Soil Res 34:273–287

Hoyle FC, Murphy DV, Fillery IRP (2006) Temperature and stubble management influence microbial CO -C evolution and gross transformation rates. Soil Biol Biochem 38:71–80

Hsieh YP (1992) Pool size and mean age of stable soil organic carbon in cropland. Soil Sci Soc Am J 56:460–464

Hubble GD (1984) The cracking clay soils: definition, distribution, nature, genesis, and use. In: MG JW, Hoult EH, So HB (eds) The properties and utilisation of cracking clay soils, Reviews in rural science 5. University of New England, Armidale, pp 3–13

Hue N (2020) Biochar for maintaining soil health. In: Giri B, Varma A (eds) Soil health, Soil biology, vol 59. Springer, Cham. https://doi.org/10.1007/978-3-030-44364-1_2

ICRISAT (1984) Annual report. International Crops Research Institute for the Semi-arid tropics, Patancheru, p 144

ICRISAT (1988) Annual report. International Crops Research Institute for the Semi-arid tropics, Patancheru, p 136

ICRISAT (1989) Annual report. International Crops Research Institute for the Semi-arid tropics, Patancheru, p 187

Ingham ER, Trofymow JA, Ames R, Hunt HW, Morley CR, Moore JC, Coleman DC (1986) Trophic interactions and nitrogen cycling in a semi-arid grassland soil. I. Seasonal dynamics of the natural populations, their interactions and effects on nitrogen cycling. J Appl Ecol 1:597–614

Insam H, Domsch KH (1988) Relationship between soil organic carbon and microbial biomass on chronosequences of reclamation sites. Microb Ecol 15:177–188

Insam H, Parkinson D, Domsch KH (1989) Influence of macro-climate on soil microbial biomass. Soil Biol Biochem 21:211–221

Inubushi K, Brookes PC, Jenkinson DS (1985) Adenosine 5′-triphosphate and adenylate energy charge in waterlogged soil. Soil Biol Biochem 21(5):733–739

Isbell RF (2016) National Committee on Soils and Terrain, The Australian soil classification, 2nd edn. CSIRO Publishing, Clayton

Jaiarree S, Chidthaisong A, Tangtham N, Polprasert C, Sarobol E, Tyler SC (2011) Soil organic carbon loss and turnover resulting from forest conversion to maize fields in Eastern Thailand. Pedosphere 21(5):581–590

Jain AN, Udayashankara TH, Lokesh KS (2014) Review on bioremediation of heavy metals with microbial isolates and amendments on soil residue. Int J Sci Res 3(8):2319–7064

Jat HS, Datta A, Choudhary M, Sharma P, Yadav A, Choudhary V, Gathala M, Jat ML, McDonald A (2019) Climate smart agriculture practices improve soil organic carbon pools, biological properties, and crop productivity in cereal-based systems of North-West India. Catena 181:104059

Jat HS, Kakraliya M, Mukhopadhyay R, Kumar S, Choudhary M, Sharma PC (2024) Conservation agriculture works as a catalyst for sustainable sodic soil reclamation and enhances crop productivity and nutrient1 use efficiency: a scientific inquiry. J Environ Manage 358:120811

Jayaraman S, Dalal RC (2022) No-till farming: prospects, challenges–productivity, soil health, and ecosystem services. Soil Res 60(6):435–441

Jayaraman S, Krishanasamy R, Savithri P (2005) Biotransfer of heavy metals in Jersey cows. Indian J Anim Sci 75(11):1257–1260

Jayaraman S, Krishnasamy R, Savithri P, Kumar BS, Behera SK, Kumar VA, Chitdeswari T, Coumar MV (2011) Effect of sewage sludge application on accumulation of Cd, Cr, Ni and Pb in forage maize *Zea mays* L. Agrochimica 55(6):332–354

Jayaraman S, Krishnasamy R, Savithri P, Coumar MV, Kumar VA, Kumar BS, Indirani R, Maragatham S, Mahimairaja S, Chitdeswari T, Sivasubramanium K (2016) Effect of sewage sludge–coir pith pellets on dry matter yield and trace metal concentration in various plant parts of forage maize. J Plant Nutr 39(11):1556–1569

Jayaraman S, Reeves S, Wang W, Heenan M, Dalal RC (2017) Impact of 47 years of no tillage and stubble retention on soil aggregation and carbon distribution in a Vertisol. Land Degrad Dev 28:1589–1602. https://doi.org/10.1002/ldr.2689

Jayaraman S, Singh RK, Prasad SN, Kumar A, Ali S, Sinha NK, Chaudhary RS, Mohanty M, Lakaria BL, Sankar M, Lal R (2018a) Effect of soil amendments and land use systems on surface cracks, soil properties and crop yield in a Vertisol. Agric Res 7:443–455

Jayaraman S, Lal R, Sinha NK, Dalal RC, Chitralekha A, Chaudhary RS, Patra AK (2018b) Cracks and potholes in vertisols: characteristics, occurrence, and management. Adv Agron 149:93–159

Jayaraman S, Sinha NK, Mohanty M, Chaudhary RS, Hati KM, Singh RK, Biswas AK, Shukla AK, Dalal RC, Patra AK (2018c) Soil hydro-thermal regimes as affected by different tillage and cropping systems in a rainfed vertisol. J Indian Soc Soil Sci 66(4):362–369

Jayaraman S, Chaudhary RS, Awanish Kumar D, Biswas AK, Sinha NK, Mohanty M, Hati KM, Jha P, Sankar M, Patra AK, Dalal RC, Chaudhari SK (2018d) Effect of contrasting tillage and cropping systems on soil aggregation, carbon pools and aggregateassociated carbon in rainfed Vertisols. Eur J Soil Sci 69:879–891. https://doi.org/10.1111/ejss.12692

Jayaraman S, Salikram M, Sinha NK, Mohanty M, Chaudhary RS, Dalal RC, Mitra NG, Blaise D, Coumar MV, Hati KM, Thakur JK (2019) Corrigendum to: conservation agriculture effects on soil properties and crop productivity in a semiarid region of India. Soil Res 57(2):200

Jayaraman S, Sinha NK, Dalal RC et al (2020) No-till farming and conservation agriculture in South Asia–issues, challenges, prospects, and benefits. Crit Rev Plant Sci 39(3):236–279

Jayaraman S, Naorem AK, Lal R, Dalal RC, Sinha NK, Patra AK, Chaudhari SK (2021) Disease-suppressive soils—beyond food production: a critical review. J Soil Sci Plant Nutr 21:1437–1465

Jayaraman S, Naorem A, Dalal RC, Sinha NK, Rao CS, Lal R, Kundu S, Prasad JV, Singh AK (2024) No-till farming and climate change mitigation: lessons learnt from long-term no-till experiments and future perspectives. Adv Agron 187:21–107. https://doi.org/10.1016/bs.agron.2024.05.005

Jenkinson DS (1988) Soil organic matter and its dynamics. In: Wild A (ed) Russell's soil conditions and plant growth. Longman, Harlow, pp 564–607

Jenkinson DS (1990) The turnover of organic carbon and nitrogen in soil. Philosophical Trans Royal Soc London. Series B: Biol l Sci 329(1255):361–368

Jenkinson DS, Ayanaba A (1977) Decomposition of carbon-14 labelled plant material under tropical conditions. Soil Sci Soc Am J 41:912–915

Jenkinson DS, Johnston AE (1977) Soil organic matter in the Hoosfield Continuous Barley Experiment. Rothamsted Experimental Station Annual Report for 1976. Pt 2, 87–101

Jenkinson DS, Ladd JN (1981) Microbial biomass in soil, measurement, and turnover. In: Paul EA, Ladd JN (eds) Soil biochemistry, vol 5. Marcel Dekker, New York, pp 415–472

Jenkinson D, Oades JM (1979) A method for measuring adenosine triphosphate in soil. Soil Biol Biochem 11(2):193–199

Jenkinson DS, Parry LC (1989) The nitrogen cycle in the Broadbalk wheat experiment: a model for the turnover of nitrogen through the soil microbial biomass. Soil Biol Biochem 21:535–541

Jenkinson DS, Rayner JH (1977) The turnover of soil organic matter in some of the Rothamsted classical experiments. Soil Sci 123:298–305

Jenkinson DS, Adams DE, Wild A (1991) Model estimates of CO emissions from soil in response to global warming. Nature (London) 351:304–306

Jenkinson DS, Harkness DD, Vance ED, Adams DE, Harrison AF (1992) Calculating net primary production and annual input of organic matter to soil from the amount and radiocarbon content of soil organic matter. Soil Biol Biochem 24:295–308

Jenny H (1941) Factors of soil formation: a system of quantitative pedology. McGraw-Hill, New York

Jenny H (1961) Derivation of state factor equations of soils and ecosystems. Proc Soil Sci Soc Am 25:385–388

Jenny H (1994) Factors of soil formation: a system of quantitative pedology. Dover Publication, INC, New York

Jha P, Hati KM, Dalal RC, Dang YP, Kopittke PM, Menzies NW (2020) Soil carbon and nitrogen dynamics in a Vertisol following 50 years of no-tillage, crop stubble retention and nitrogen fertilization. Geoderma 358:113996

Jha P, Lakaria BL, Vishwakarma AK, Wanjari RH, Mohanty M, Sinha NK, Somasundaram J, Dheri GS, Dwivedi AK, Sharma RP, Singh M, Dalal RC, Biswas AK, Patra AK, Chaudhari SK (2021) Modeling the organic carbon dynamics in long-term fertilizer experiments of India using the Rothamsted carbon model. Ecol Model 450:109562

Jiao W, Chen W, Chang AC, Page AL (2012) Environmental risks of trace elements associated with long-term phosphate fertilizers applications: a review. Environ Pollut 168:44–53

Jie C, Jing-Zhang C, Man-Zhi T, Zi-tong G (2002) Soil degradation: a global problem endangering sustainable development. J Geogr Sci 12:243–252

Joergensen RG, Emmerling C (2006) Methods for evaluating human impact on soil microorganisms based on their activity, biomass, and diversity in agricultural soils. J Plant Nutr Soil Sci 169:295–309. https://doi.org/10.1002/jpln.200521941

Joergensen RG, Hemkemeyer M, Beule L, Iskakova J, Oskonbaeva Z, Rummel PS, Schwalb SA, Wichern F (2024) A hitchhiker's guide: estimates of microbial biomass and microbial gene abundance in soil. Biol Fertil Soils 19:1–4

Johnston AM, Bruulsema TW (2014) 4R nutrient stewardship for improved nutrient use efficiency. Procedia Eng 83:365–370

Jones AR, Dalal RC, Gupta VVSR, Schmidt S, Allen DE, Jacobsen GE, Bird M, Grandy AS, Sanderman J (2023) Molecular complexity and diversity of persistent soil organic matter. Soil Biol Biochem 184:109061

Jorge-Silva ML, Ferraz JF, Rodrigues ML, Ricardo CP (1989) Effect of water availability on growth of Fusarium oxysporum f. sp. melonis and on host-parasite interactions. Myco Res 92 (2):157–161

Juo AS, Manu A (1996) Chemical dynamics in slash-and-burn agriculture. Agric Ecosyst Environ 58(1):49–60

Kabata-Pendias A (2000) Trace elements in soils and plants. CRC Press

Kadomura J, Yamamoto H (1978) Man-induced erosion: the rate of erosion and an example from Okinawa Island, southern Japan. J Geogr 87:1–15

Kallenbach CM, Frey SD, Grandy AS (2016) Direct evidence for microbial-derived soil organic matter formation and its ecophysiological controls. Nature commun 7(1): 13630.

Kakraliya SK, Singh U, Bohra A, Choudhary KK, Kumar S, Meena RS, Jat JL (2018) Nitrogen and legumes: a meta-analysis. Legumes for soil health sustainable management, In, pp 277–314

Kang BT, Juo ASR (1986) Effect of forest clearing on soil chemical properties and crop performance. In: Lal R, Sanchez PA, Cummings RW (eds) Land clearing and development in the tropics. Balkema AA, Rotterdam, pp 383–394

Karl DM (1980) Cellular nucleotide measurements and applications in microbial ecology. Microbiol Rev 44:739–796

Katayama A, Bhula R, Burns GR, Carazo E, Felsot A, Hamilton D, Harris C, Kim YH, Kleter G, Koedel W, Linders J (2010) Bioavailability of xenobiotics in the soil environment. In: Whitacre D (ed) Reviews of environmental contamination and toxicology, vol 203. Springer, New York, pp 1–86

Katyal JC, Sharma BD (1991) DTPA-extractable and total Zn, Cu, Mn, and Fe in Indian soils and their association with some soil properties. Geoderma 49(1–2):165–179

Kaur R, Page KL, Dalal RC, Menzies NW, Dang YP (2022) Can molecular genetic techniques improve our understanding of the role the microbial biomass plays in soil aggregation? Eur J Soil Sci 73(5):e13307

Kawalec A (1973) World distribution of acid sulfate soils: reference and map. In: Dost H (ed) Acid sulfate soils, vol I. International Institute for Land Reciamation and Improvement, Wageningen, pp 292–295

Kay B, Vandenbygaart B (2002) Conservation tillage and depth stratification of porosity and soil organic matter. Soil Tillage Res 66:107–118. https://doi.org/10.1016/S0167-1987(02)00019-3

Keating BA, Carberry PS, Hammer GL, Probert ME, Robertson MJ, Holzworth D, Huth NI, Hargreaves JN, Meinke HZ, McLean G (2003) An overview of APSIM, a model designed for farming systems simulation. Eur J Agron 18(3–4):267–288

Keeling CD (1979) The Suess effect: ^{13}Carbon-^{14}Carbon interrelations. Environ Int 2(4–6):229–300

Keeney DR (1982) Nitrogen availability indices. In: Page AL et al (eds) Methods of soil analysis Part 2. American Society of Agronomy, Madison, pp 711–733

Keesstra S, Mol G, de Leeuw J, Okx J, Molenaar C, de Cleen M, Visser S (2018) Soil-related sustainable development goals: four concepts to make land degradation neutrality and restoration work. Land 7:133

Kelley BC, Tuovinen OH (1988) Microbiological oxidations of minerals in mine tailings. In: Chemistry and biology of solid waste: dredged material and mine tailings. Springer, Berlin/Heidelberg, pp 33–53

Kichamu-Wachira E, Xu XZ, Reardon-Smith K, Biggs D, Wachira G, Omidvar N (2021) Effects of climate-smart agricultural practices on crop yields, soil carbon, and nitrogen pools in Africa: a meta-analysis. J Soils Sediments 21:1587–1597

Kirkby CA, Kirkegaard JA, Richardson AE, Wade LJ, Blanchard C, Batten G (2011) Stable soil organic matter: a comparison of C:N:P:S ratio in Australian and other world soils. Geoderma 163:197–208

King AE, Congreves KA, Deen B, Dunfield KE, Simpson MJ, Voroney RP, Wagner-Riddle C (2020) Crop rotations differ in soil carbon stabilization efficiency, but the response to quality of structural plant inputs is ambiguous. Plant Soil 457: 207–224

Kiss S, Drăgan-Bularda M, Rădulescu D (1975) Biological significance of enzymes accumulated in soil. Adv Agron 27:25–87

Kleber M, Eusterhues K, Keiluweit M, Mikutta C, Mikutta R, Nico PS (2015) Mineral-organic associations: formation, properties, and relevance in soil environments. Adv Agron 130:3–140

Klose S, Tabatabai MA (2000) Urease activity of microbial biomass in soils as affected by cropping systems. Biol Fertil Soils 31:191–199

Knisel WG (1980) CREAMS: a field-scale model for chemicals, runoff, and erosion from agricultural management systems. USDA conservation Research Report No. 26. USDA-ARS, Washington, DC

Knowles R (1982) Denitrification. Microbiol Rev 46(1):43–70

Knowles R (2005) Denitrifiers associated with methanotrophs and their potential impact on the nitrogen cycle. Ecol Eng 24(5):441–446

Kommedahl T, Windels CE (1979) Fungi: pathogen or host dominance in disease. In: Developments in agricultural and managed forest ecology 5. Elsevier, pp 1–103

Kommedahl T, Windels CE, Stucker RE (1979) Occurrence of Fusarium species in roots and stalks of symptomless corn plants during the growing season. Phytopathology 69(9):961–966

Kopittke PM, Menzies NW (2007) A review of the use of the basic cation saturation ratio and the 'ideal' soil. Soil Sci Soc Am J 71(2):259–265

Kopittke PM, Dalal RC, Menzies NW (2016) Sulfur dynamics in sub-tropical soils of Australia as influenced by long-term cultivation. Plant Soil 402:211–219

Kopittke PM, Dalal RC, Finn D, Menzies NW (2017) Global changes in soil stocks of carbon, nitrogen, phosphorus, and sulphur as influenced by long-term agricultural production. Glob Change Biol 23(6):2509–2519

Kopittke PM, Hernandez-Soriano MC, Dalal RC, Finn D, Menzies NW, Hoeschen C, Mueller CW (2018) Nitrogen-rich microbial products provide new organo-mineral interactions for the stabilization of soil organic matter. Glob Chang Biol 24:1762–1770

Kopittke PM, Dalal RC, Hoeschen C, Li C, Menzies NW, Mueller CW (2020) Soil organic matter is stabilized by organo-mineral associations through two key processes: the role of the carbon to nitrogen ration. Geoderma 357:113974

Körschens M (2005) Global and regional importance of long-term field experiments. Arch Agron Soil Sci 51:111–117

Kotroczo Z, Veres Z, Fekete I, Krakomperger Z, Tóth JA, Lajtha K, Tóthmérész B (2014) Soil enzyme activity in response to long-term organic matter manipulation. Soil Biol Biochem 70:237–243

Kraamwinkel CT, Beaulieu A, Dias T, Howison RA (2021) Planetary limits to soil degradation. Commun Earth Environ 2(1):249

Krishnamoorthy RV (1989) Factors affecting the surface cast production by some earthworms of Indian tropics. Proc Anim Sci 98:431–445

Krishnamoorthy CS, Chattopadya RSBP (1966) Safflower for better utilisation of shallow and medium black cotton soils of Bellary areas. Annals Arid Zone 5:44–48

Kronvang B, Vagstad N, Behrendt H, Bøgestrand J, Larsen SE (2007) Phosphorus losses at the catchment scale within Europe: an overview. Soil Use Manag 23:104–116

Krug EC, Frinck CR (1983) Acid rain on acid soil: a new perspective. Science 221:520–525

Kuhlbusch TA, Lobert JM, Crutzen PJ, Warneck P (1991) Molecular nitrogen emissions from denitrification during biomass burning. Nature (London) 351:135–137

Kwari JD, Batey T (1991) Effect of heating on phosphate sorption and availability in some north-east Nigerian soils. J Soil Sci 42:381–388

Kyuma K, Tulaphitak T, Pairintra C (1985) Changes in soil fertility and tilth under shifting cultivation 1. General description of soil and effect of burning on the soil characteristics. Soil Sci Plant Nutr 31:227–238

Ladd JN, Parssons JW, Amato M (1977) Studies of N immobilisation and mineralisation in calcareous soils: 1. Distribution of immobilised N amongst soil fractions of different particle size and density. Soil Biol Biochem 9:309–319

Lakaria BL, Singh M, Sammi Reddy K, Biswas AK, Jha P, Chaudhary RS, Singh AB, Subba Rao A (2012) Carbon addition and storage under integrated nutrient management in soybean–wheat cropping sequence in a Vertisol of central India. Natl Acad Sci Lett 35:131–137

Lal R (1978) Influence of within- and between-row mulching on soil temperature, soil moisture, root development and yield of maize in a tropical soil. Field Crops Res 1:127–139

Lal R (1980) Losses of plant nutrients in runoff and eroded soil. In: Rosswall T (ed) Nitrogen cycling in West African ecosystems. Reklan and Katalogtryck, Uppsala, pp 31–38

Lal R (1983) Effects of slope length on runoff from an alfisol in western Nigeria. Geoderma 31:185–193

Lal R (1984) Mechanized tillage systems effects on soil erosion from an Alfisols in watersheds cropped to maize. Soil Tillage Res 4:349–360

Lal R (1985) Soil erosion and sediment transport research in tropical Africa. Hydrol Sci J 30 (2):239–256. https://doi.org/10.1080/02626668509490987

Lal R (1988) Effects of macrofauna on soil properties in tropical ecosystems. Agric Ecosyst Environ 24:101–116

Lal R (1989) Conservation tillage for sustainable agriculture: tropics versus temperate environments. Adv Agron 42:85–191

Lal R (1990a) Soil erosion and land degradation: the global risks. Adv Soil Sci 11:129–172

Lal R (1990b) Soil erosion in the tropics: principles and management. McGraw Hill, New York

Lal R (1991) Soil conservation and biodiversity. In: Hawksworth DL (ed) The biodiversity of microorganisms and invertebrates: its role in sustainable agriculture. CAB Int'l, Wallingford, pp 89–104

Lal R (1996) Deforestation and land-use effects on soil degradation and rehabilitation in western Nigeria. III. Runoff, soil erosion and nutrient loss. Land Degrad Dev 7(2):99–119

Lal R (1998a) Soil erosion impact on agronomic productivity and environment quality. Crit Rev Plant Sci 17(4):319–464

Lal R (1998b) Soil degradative effects of slope length and tillage methods on alfisols in western Nigeria. III. Soil physical properties. Land Degrad Dev 8(4):325–342

Lal R (2000) Physical management of soils of the tropics: priorities for the 21st century. Soil Science 165(3):191–107

Lal R (2001) Soil degradation by erosion. Land degradation & development 12(6):519–539

Lal R (2002) Soil carbon dynamics in cropland and rangeland. Environ Pollut 116:353–362
Lal R (2003) Soil erosion and the global carbon budget. Environ Int 29(4):437–450
Lal R (2008) Promise and limitations of soils to minimise climate change. J Soil Water Conserv 63:113A–118A
Lal R (2009) Soil degradation as a reason for inadequate human nutrition. Food Sec 1:45–57
Lal R (2015a) Cover cropping and the "4 per Thousand" proposal. J Soil Water Conserv 70 (6):141A
Lal R (2015b) Restoring soil quality to mitigate soil degradation. Sustain For 7(5):5875–5895
Lal R (2016) Soil health and carbon management. Food Energy Secur 5:212–222. https://doi.org/10.1002/fes3.96
Lal R (2018) Digging deeper: a holistic perspective of factors affecting soil organic carbon sequestration in agroecosystems. Glob Change Biol 24(8):3285–3301
Lal R (2019) Land-use impacts on soil physical properties of an Alfisol in Western Nigeria. In: Soil Degradation and Restoration in Africa, pp 60–70
Lal R (2020) Managing soil quality for humanity and the planet. Front Agric Sci Eng 7(3):251–253
Lal R (2020a) Soil organic matter content and crop yield. J Soil Water Cons 75:27A–32A. https://doi.org/10.2489/jswc.75.2.27A
Lal R (2020b) Soil organic matter and water retention. Agron J 12(5):3265–3277
Lal R (2021a) Feeding the world and returning half of the agricultural land back to nature. J Soil Water Conserv 76(4):75A–78A
Lal R (2021b) Negative emission farming. J Soil Water Conserv 76(3):61A–64A
Lal R (2023) Farming systems to return land for nature: it's all about soil health and re-carbonization of the terrestrial biosphere. Farm Syst 1(1):100002
Lal R (2024) A historic landmark for the role of soil health in climate and food security: "soil health" in COP28 UAE declaration. J Soil Water Conserv 79(3):48A–52A
Lal R, Cummings DJ (1979) Clearing a tropical forest I. Effects on soil and micro-climate. Field Crops Res 2:91–07
Lal R, Kang BT (1982) Management of organic matter in soils of the tropics and subtropics. 12th ISSS Congr New Delhi Symp 1:152–178
Lal R, Moldenhauer WC (1987) Effects of soil erosion on crop productivity. Crit Rev Plant Sci 5 (4):303–367
Lal R, Shukla MK (2004) Principles of soil physics. CRC Press, Boca Raton
Lal R, Stewart BA (1990) Soil degradation: a global threat. Adv Soil Sci 11:xiii–xvii
Lal R, Stewart BA (eds) (2018) Soil nitrogen uses and environmental impacts. CRC Press, Boca Raton
Lal R, Hall GF, Miller FP (1989) Soil degradation: basic processes. Land Degrad Dev 1(1):51–69
Lal R, Bouma J, Brevik E, Dawson L, Field D, Glaser B, Hatano R, Hartemink AE, Kosaki T, Lascelles B, Monger C, Muggler C, Ndzana GM, Norra S, Pan X, Paradelo R, Reyes-Sanchez LB, Sanden T, Singh BR, Spiegel H, Yanai J, Zhang J (2021) Soils and sustainable development goals of the United Nations: an International Union of Soil Sciences perspective. Geoderma Reg 25:e00398
Lambin EF, Meyfroidt P (2021) Global land use change, economic globalization, and the looming land scarcity. Proc Natl Adv Sci 108(9):3465–3472
Langridge P (2022) Micronutrient toxicity and deficiency. In: Wheat improvement: food security in a changing climate. Springer, Cham, pp 433–449
Larson WE, Fenton TE, Skidmore EL, Benbrook CM (1985) Effects of soil erosion on soil properties as related to crop productivity and classification. In: Follett RF, Stewart BA (eds) Soil erosion and crop productivity. American Society of Agronomy, Madison, pp 190–212
Latham M (1987) IBSRAM's soil management approach, vol 7. Land Development and Management of Acid Soils in Africa II, Africaland, p 11
Lathwell DJ, Bouldin DR (1981) Soil organic matter and soil nitrogen behaviour in cropped soils. Trop Agric Trinidad 58:341–348

Laub M, Blagodatsky S, van de Broek M, Schlichenenmaier S, Kunlanit B, Six J, Vityakon P, Cadisch G (2024) SAMM version 1.0: a numerical model for microbial-mediated soil aggregate formation. Geosci Model Dev 17:931–956
Lavallee JM, Soong JL, Cotrufo MF (2020) Conceptualizing soil organic matter into particulate and mineralassociated forms to address global change in the 21st century. Global Change Biol 26 (1):261–273
Lavelle P (1988) Earthworm activities and the soil system. Biol Fert Soils 6:237–251
Lavelle P, Blanchart E, Martin A, Martin S, Barois I, Toutain F, Spain A, Schaefer R (1993) A hierarchical model for decomposition in terrestrial ecosystems. Application to soils in the humid tropics. Biotropica 25:130–150
Lavelle P, Barros E, Blanchart E, Brown G, Desjardins T, Mariani L, Rossi JP (2001) SOM management in the tropics: why feeding the soil macrofauna? In: Managing organic matter in tropical soils: scope and limitations: proceedings of a workshop organized by the Center for Development Research at the University of Bonn (ZEF Bonn)—Germany. Springer, Dordrecht, pp 53–61
Lavelle P, Charpentier F, Villenave C, Rossi JP, Derouard L, Pashanasi B, André J, Ponge JF, Bernier N (2004) Effects of earthworms on soil organic matter and nutrient dynamics at a landscape scale over decades. In: Earthworm ecology. CRC Press, pp 145–160
Lee KE (1985) Earthworms: their ecology and relationships with soils and land use. Academic, New York
Lee KE, Foster RC (1991) Soil fauna and soil structure. Aust J Soil Res 29:745–775
Lee KE, Pankhurst CE (1992) Soil organisms and sustainable productivity. Aust J Soil Res 30:855–892
Lee KE, Wood TG (1971) Termites and soils. Academic, London, p 251
Legout C, Leguédois S, Le Bissonnais Y (2005) Aggregate breakdown dynamics under rainfall compared with aggregate stability measurements. Eur J Soil Sci 56(2):225–238
Leininger S, Urich T, Schloter M, Schwark L, Qi J, Nicol GW, Prosser JI, Schuster SC, Schleper C (2006) Archaea predominate among ammonia-oxidizing prokaryotes in soils. Nature 442:806–809
Lenvain JS, Chinene VRN, Gill KS, Pauwelyn PLL (1987) Aspects of compaction, water transmissibility and erosion on a kaolinitic clay soil. In: Latham M, Ahn P (eds) Land development and management of acid soils in Africa. International Board for Soil Research and Management, Bangkok, pp 173–185
Leogrande R, Vitti C (2019) Use of organic amendments to reclaim saline and sodic soils: a review. Arid Land Res Manag 33(1):1–21
Lepcha NT, Devi NB (2020) Effect of land use, season, and soil depth on soil microbial biomass carbon of Eastern Himalayas. Ecol Process 9:1–4
Lesack LFW, Melack JM (1991) The deposition, composition, and potential sources of major ionic solutes in rain of the central Amazon basin. Water Resour Res 27:2953–2977
Letey J (1985) Relationship between soil physical properties and crop production. Adv Soil Sci 1:277–294
Lev H, He P, Zhao S (2024) Optimised nitrogen fertilization promoted soil organic carbon accumulation by increasing microbial necro mass carbon in potato continuous cropping field. Agronomy 14:307
Levy GJ, van der Watt HVH (1988) Effects of clay mineralogy and soil sodicity on soil infiltration rate. S Afr J Plant Soil 5:92–96
Lewis T, Verstraten HB, Wehr BJ, Swift S, Tindale N, Menzies NW, Dalal RC, Bryant P, Francis B, Smith TE (2019) Reforestation of agricultural land in the tropics: the relative contribution of soil, living biomass and debris pools to carbon sequestration. Sci Total Environ 649:1502–1513
Li C, Folkring S, Folkring TA (1992) A model of nitrous oxide evolution from soil driven by rainfall events: 1. Model structure and sensitivity. J Geophysical Res 97:9759–9776
Li H, Zhu H, Wei X, Liu B, Shao M (2021a) Soil erosion leads to degradation of hydraulic properties in the agricultural region of Northeast China. Agric Ecosyst Environ 314:107388

Li J, Fu B, Feng X, Dargusch P, Liu J, Chen W, Wang J (2021b) Topsoil carbon selective transport in an eroding soil landscape with vegetation restoration. Land Degrad Dev 32(5):2061–2073
Li Y, Liu W, Feng Q, Zhu M, Yang L, Zhang J (2022) Effects of land use and land cover change on soil organic carbon storage in the Hexi regions, Northwest China. J Environ Manage 312:114911
Li Z, Kravchenko N, Cupples A, Guber AK, Robertson GP, Blagodatskaya E (2024) Compositionand metabolism of microbial communities in soil pores. Nat Commun 15:3578
Liao Z, Fan J, Lai Z, Bai Z, Wang H, Cheng M, Zhang F, Li Z (2023) Response network and regulatory measures of plant-soilrhizosphere environment to drought stress. Adv Agron 180:93–196
Linden DR, Hendrix PF, Coleman DC, van Vliet PC (1994) Faunal indicators of soil quality. In: Doran JW et al (eds) Defining soil quality for a sustainable environment, vol 35, pp 91–106
Linderman RG (1997) Vesicular-arbuscular mycorrhizal (VAM) fungi. In: Plant Relationships Part B. Springer, Berlin/Heidelberg, pp 117–128
Lindsay WL (1991) Inorganic equilibria affecting micronutrients in soils. Micronutr Agric 4:89–112
Lindsay WL, Norvell WA (1978) Development of a DTPA soil test for zinc, iron, manganese, and copper. Soil Sci Soc Am J 42:421–428
Lindstrom MJ, Schumacher TE, Lemme GD, Gollany HM (1986) Soil characteristics of a mollisol and corn (*Zea mays* L) growth 20 years after topsoil removal. Soil Tillage Res 7(1–2):51–62
Lipsett J, Dan PR (1983) Wheat, Australia's hidden mineral export. J Aust Inst Agric Sci 49:81–91
Littleboy M, Silburn DM, Freebairn DM, Woodruff DR, Hammer GL (1989) Perfect, a computer simulation model of productivity, erosion, runoff functions to evaluate conservation techniques. Queensland Department of Primary Industries, Brisbane
Liu J, You L, Amini M, Obersteiner M, Herrero M, Zehnder AJ, Yang H (2010) A high-resolution assessment on global nitrogen flows in cropland. Proc Nat Acad Sci 107(17):8035–8040
Lo KFA (1989) Erosion-productivity interrelationships for tropical soils, soils and fertilisers in Taiwan 19–30. Society of Soil and Fertilizer Science Report China
Loader LP, White PJ (1979) Sulphur deficiency in pastures on the Darling Downs. Queensland Agric J 105:45–48
Lobo D, Lozano Z, Delgado F (2005) Water erosion risk assessment and impact on productivity of a Venezuelan soil. Catena 64(2–3):297–306
Loch RJ (1989) Aggregate breakdown under rain: its measurement and interpretation. University of New England, Armidale, PhD Disser
Loch RJ (1994) Effects of fallow management and cropping history on aggregate breakdown under rainfall wetting for a range of Queensland soils. Soil Res 32(5):1125–1139
Loch RJ, Foley JL (1994) Measurement of aggregate breakdown under rain: comparison with tests of water stability and relationships with field measurements of infiltration. Aust J Soil Res 32:701–720
Logan TJ (1990) Chemical degradation of soil. Advances in soil science, soil degradation 11:187–221
Logan TJ (1992) Reclamation of chemically degraded soil. Adv Soil Sci 17:13–36
Logan TJ, Chaney RL (1983) Utilisation of municipal wastewater and sludge on land-metals. In: Page AL, Gleason TL, Smith JE Jr, Iskander IK, Sommers LE (eds) Proceeding of the Workshop on Utilization of Municipal Wastewater and Sludge on Land. Univ California, Riverside, pp 235–326
Loll MJ, Bollag JM (1983) Protein transformation in soil. Adv Agron 36:351–382
Long SP, Moya EG, Imbabma SK, Kamnalrut A, Piedade MTF, Scurlock JMO, Shen YK, Hall DO (1989) Primary productivity of natural grass ecosystems of the tropics: a reappraisal. Plant Soil 115:155–166
Lopes AS, Cox FR (1977) A survey of the fertility status of surface soils under 'Cerrads' vegetation in Brazil. Soil Sci Soc Am J 41:742–747

Lopes AS, Smyth TJ, Curi N (1987) The need for a soil fertility reference base and nutrient dynamics studies. In: Sanchez PA, Stoner ER, Pushparajah E (eds) Management of acid tropical soils for sustainable agriculture. International Board for Soil Research and Management, Bangkok, pp 147–165

Lorenz K, Lal R (eds) (2018) Carbon sequestration in agricultural ecosystems. Springer

Lorenz K, Lal R, Ehlers K (2019) Soil organic carbon stock as an indicator for monitoring land and soil degradation in relation to United Nations' Sustainable Development Goals. Land Degrad Dev 30(7):824–838. https://doi.org/10.1002/ldr.3270

Loveday J (1976) Relative significance of electrolyte and cation exchange effects when gypsum is applied to a sodic clay soil. Aust J Soil Res 14:361–371

Low AJ (1972) The effect of cultivation on the structure and other physical characteristics of grassland and arable soils (1945–1970). J Soil Sci 23:363–380

Lugato E, Lavallee JM, Haddix ML, Panagos P, Cotrufo MF (2021) Different climate sensitivity of particulate and mineralassociated soil organic matter. Nat Geosci 14:295–300

Lungu OIM (1987) Monitoring soil fertility. In: Latham M, Ahm P (eds) Land development and management of acid soils in Africa. International Board for Soil Research and Management, Bangkok, pp 161–171

Luvia A, Obiero J, Omoto C (2022) Soil loss assessment using the Universal Soil Loss Equation (RUSLE) model. Appl Environ Soil Sci 2022:2122554

Lyles L (1977) Wind erosion: processes and effect on soil productivity. Trans Am Soc Agric Eng 20:880–884

Maas EV (1986) Salt tolerance of plants. In: Christie BR (ed) Handbook of plant science in agriculture. CRC Press Inc, Baton Raton

Maas EV, Hoffman GJ (1977) Crop salt tolerance-current assessment. J IRRI Div Am Soc Civil Eng 103:115–134

MacDonald GK, Bennett EM, Potter PA, Raman Kutty N (2011) Agronomic phosphorus imbalances across the world's croplands. Proc Natl Acad Sci 108(7):3086–3091

Madalcho AB, Sido MY (2015) Performance of in-situ rainwater conservation tillage techniques and inorganic fertilizer practices on sorghum production at Ethiopia Somali Region (Kurdha Metan district). Int J Agric Sci Res 4(5):98–108

Majer JD (1983) Ants: bio-indicators of minesite rehabilitation, land-use, and land conservation. Environmental management 7:375–383

Mandal KG, Misra AK, Hati KM, Bandyopadhyay KK, Ghosh PK, Mohanty M (2004) Rice residue management options and effects on soil properties and crop productivity. Food Agric Environ 2:224–231

Maqsood MA, Hussain S, Aziz T, Ashraf M (2013) Sustainable agriculture through integrated soil fertility management on degraded lands, developments in soil salinity assessment and reclamation: innovative thinking and use of marginal soil and water resources in irrigated agriculture, pp 759–768

Maranguit D, Guillaume T, Kuzyakov Y (2017) Land-use change affects phosphorus fractions in highly weathered tropical soils. Catena 149:385–393

Mariotti A, Balesdent J (1990) 13C natural abundance as a tracer of soil organic matter turnover and paleoenvironment dynamics. Chemical geology 84(1-4):217–219

Marschner P, Rengel Z (eds) (2007) Nutrient cycling in terrestrial ecosystems. Springer, Berlin/Heidelberg, pp 1–397

Marsh B (1981) Wind erosion in relation to tillage systems. In: Proceeding of a national workshop on Tillage Systems for Crop Production. Roseworthy, South Australia, pp 822–825

Marshall TJ (1962) The nature, development, and significance of soil structure. In: Transactions of joint meeting of commissions IV and V. International Society of Soil Science, New Zealand, pp 243–257

Marshall JK (1973) Drought, land use and soil erosion. In: Lovette JV (ed) The environmental, economic and social significance of drought. Angus and Robertson, Sydney

Martel YA, Paul EA (1974) Effects of cultivation on the organic matter of grassland soils as determined by fractionation and radiocarbon dating. Can J Soil Sci 54:419–426

Martens R (1985) Estimation of the adenylate energy charge in unamended and amended agricultural soils. Soil Biol Biochem 17:765–772

Martin AE, Cox JE (1956a) Nitrogen studies on black soils from Darling Downs, Queensland 1. Seasonal variations in moisture and mineral nitrogen fractions. Aust J Agric Res 7:69–183

Martin AE, Cox JE (1956b) Nitrogen studies on black soils from Darling Downs, Queensland 1. Seasonal variations in moisture and critical limits in Vertisols. Indian J Fertil 15(7):748–753

Martin JP, Martin WP, Page JB, Raney WA, De Ment JG (1955) Soil aggregation. Adv Agron 7:1–37

Mathers N, Dalal RC, Moody P, Mara Seni T (2006) Carbon sequestration – a case study from South Burnett. In: Jones CE, Clarke K (eds) Proceedings 'managing the carbon cycle', Kingaroy Forum, 25–26 Oct. 2006. Carbon for Life Inc, pp 31–34

Mbagwu JSC (1988) Physico-chemical properties and productivity of an Ultisols in Nigeria as affected by long-term erosion. Pedology 38:137–154

Mbagwu JSC (1989) Effects of organic amendments on some physical properties of a tropical Ultisol. Biol Wastes 28(1):1–13

McBeath TM, Grant CD, Murray RS, Hildenborough DJ (2010) Effects of subsoil amendments on soil physical properties, crop response, and soil water quality in a dry year. Aust J Soil Res 48:140–149

McBride MB (1989) Reactions controlling heavy metal solubility in soils. Adv Soil Sci 10:1–56

McCown RL (1983) An agroecological approach to management of SAT Alfisols. In: Proceedings of the Consultants' Workshop on the State of the Art and Management Alternatives for Optimising the Productivity of SAT Alfisols and Related Soils, pp 175–185

McGill WB (1996) Review and classification of ten soil organic matter (SOM) models. In: Evaluation of soil organic matter models: Using existing long-term datasets. Springer, Berlin/Heidelberg, pp 111–132

McGrath SP, Brookes PC, Giller KE (1998) Effects of potentially toxic metals in soil derived from past applications of sewage sludge on nitrogen fixation by Trifolium repens L. Soil Biol Biochem 20(4):415–424

McKeague JM (1971) Organic matter in particle-size and specific gravity fractions of some Ah horizons. Can J Soil Sci 51:499–505

McLaren AD (1975) Soil as a system of humus and clay immobilized enzymes. Chem Scr 8:97–99

McNeal BL, Coleman NT (1966) Effect of solution composition on soil hydraulic conductivity. Soil Sci Soc of Am J 30(3):308–312

McNeal BL, Norvell WA, Coleman NT (1966) Effect of solution composition on the swelling of extracted soil clays. Soil Sci Soc of Am J 30(3):313–317

Meena RS, Das A, Yadav GS, Lal R (eds) (2018) Legumes for soil health and sustainable management. Springer, Singapore, p 541

Meentemeyer V, Gardner J, Box EO (1985) World patterns and amounts of detrital soil carbon. Earth Surface Processes and Landforms 10(6):557–567

Mendes R, Garbeva P, Raaijmakers JM (2013) The rhizosphere microbiome: significance of plant beneficial, plant pathogenic, and human pathogenic microorganisms. FEMS Microbiol Rev 37:634–663

Mentges MI, Reichert JM, Rodrigues MF, Awe GO, Mentges LR (2016) Capacity and intensity soil aeration properties affected by granulometry, moisture, and structure in no-tillage soils. Geoderma 263:47–59

Menzel RG (1980) Enrichment ratios for water quality modelling. In: Knisel WG (ed) Creams. A field-scale model for chemicals, runoff, and erosion from agricultural management systems. USDA, Conservation Research Report No. 26, Washington D.C., pp 486–492

Menzies N, Bell M, Kopittke P (2019) Subsoil constraints: the many challenges of sodium, chloride, and salinity and what you can do about them. viewed 02 April May 2024 https://grdc.com.au/resources-and-publications/grdc-update-papers/tab-content/grdcupdatepapers/2019/02/subsoil-constraints-the-many-challenges-of-sodium,-chloride,-and-salinity-and-what-you-can-do-aboutthem

Meredith JA, Inserra RN, de Fernandez DM (1983) Parasitism of Rotylenchulus reniformis on soybean root Rhizobium nodules in Venezuela. J Nematol 15(2):211

Metherall AK, Harding LA, Cole CV, Parton WJ (1993) CENTURY SOM Model Environment, Technical Documentation, Agroecosystem Version 4.0. GPSR Unit, Technical Report No. 4. USDA-ARS. Fort Collins, CO

Meyer LD, Bauer A, Heil RD (1985) Experimental approaches for quantifying the effect of soil erosion on productivity. In: Follett RF, Stewart BA (eds) Soil erosion and crop productivity. American Society of Agronomy, Madison, pp 214–233

Mishra S et al (2019) Heavy metal contamination: an alarming threat to environment and human health. In: Sobti R, Arora N, Kothari R (eds) Environmental biotechnology for sustainable future. Springer, Singapore. https://doi.org/10.1007/978-981-10-7284-0_5

Misra RV, Maheshwari S (1988) Integrated nutrient management in drylands. In: Willinger PW, Sneed TV, Jordon WR, Jensen R (eds) Changes in dryland agriculture. A global perspective. Proceedings of the international conference on Dryland Farming, Amarillo, Bushland, Texas, pp 371–374

Mohanty M, Sinha NK, Somasundaram J, McDermid SS, Patra AK, Singh M, Dwivedi AK, Reddy KS, Rao CS, Prabhakar M, Hati KM (2020) Soil carbon sequestration potential in a Vertisol in central India-results from a 43-year long-term experiment and APSIM modelling. Agric Syst 184:102906

Montagnon F, Buschbacher R (1989) Nitrification rates in two undisturbed tropical rain forests and three slash-and-burn sites of the Venezuelan Amazon. Biotrophic 21:9–14

Montanarella L, Pennock D, McKenzie N, Alavipanah SK, Alegre J, Alshankiti A, Arrouays D, Singh Aulakh M, Badraoui M, dos Santos Baptista Costa I, Black H, Camps Arbestain M, Chude V, EAEM E-S, Espinosa-Victoria D, Hempel J, Henriquez CR, Young Hong S, Krasilnikov P, Mamo T, Mendonça-Santos M d L, Sobocka J, Taboada M, Vijarnsorn P, Yagi K, Yemefack M, Zhang G-L (2015) The status of the world's soil resources (technical summary). Food and Agriculture Organization of the United Nations, Rome

Montgomery DR (2007) Soil erosion and agricultural sustainability. Proc Natl Acad Sci 104 (33):13268–13272

Moore AW, Russell JS (1972) Factors affecting dehydrogenase activity as an index of soil fertility. Plant and Soil 1:675–682

Moraes VR, Brito OR, Fioretto RA, Moreira A (2016) Changes in chemical properties of an oxisol due to gypsum application. Comm Soil Sci Plant Anal 47(5):571–580

Moraghan JT, Mascagni HJ Jr (1991) Environmental and soil factors affecting micronutrient deficiencies and toxicities. Micronutrients in agriculture 4:371–425

Moraghan JT, Rego TJ, Buresh RJ, Vlach PLG, Burford JR, Singh S, Saharawat KL (1984) Labelled nitrogen fertiliser research with urea in the semi-arid tropics II. Field studies on a Vertisol. Plant Soil 80:21–33

Moreau R (1978) Effect of mechanical loosening and water infiltration on the structural stability of a ferality soil in central Ivory Coast. Pedology 1654:413–424

Mosier A, Schimel D, Valentine D, Bronson K, Parton W (1991) Methane and nitrous oxide fluxes in native, fertilized, and cultivated grassland. Nature (London) 350:330–332

Mosier S, Cordova S, Robertson GP (2021) Restoring soil fertility on degraded lands to meet food, fuel, and climate security needs via perennialization. Front Sustain Food Syst 5:706142

Mueller-Harvey IA, Juo AS, Wild A (1985) Soil organic C, N, S and P after forest clearance in Nigeria: mineralization rates and spatial variability. J Soil Sci 36(4):585–591

Muhammad I, Wang J, Sanju UM, Zhang S, Zhao F, Khan A (2021) Cover cropping enhances soil microbial biomass and affects microbial community structure: a meta-analysis. Geoderma 381:114696

Mullins CE, MacLeod DA, Northcote KH, Tisdall JM, Young IM (1990) Hard setting soils: behaviour, occurrence, and management. Adv Soil Sci 11:37–108

Mulvaney RL, Bremner JM (1978) Use of p-benzoquinone and hydroquinone for retardation of urea hydrolysis in soils. Soil Biol Biochem 10(4):297–302

Mulvaney R, Khan S, Mulvaney C (1997) Nitrogen fertilizers promote denitrification. Bio Fertil Soils 24:211–220. https://doi.org/10.1007/s003740050233

Musa MM, Mukhtar NO (1969) Enzymatic activity of a soil profile in the Sudan Gezira. Plant Soil 30:153–156

NAAS (2010) Degraded and wastelands of India: status and spatial distribution. Directorate of information and publications of agriculture, Indian Council of Agricultural Research, New Delhi, pp 01–158

Naidu R, Sumner ME (1998) Sodic soils: distribution, properties, management, and environmental consequences. Oxford University Press

Nakamura K, Harter T, Hirono Y, Horino H, Mitsuo T (2004) Assessment of root zone nitrogen leaching as affected by irrigation and nutrient management practices. Vadose Zone J 3(4):1353–1366

Nangju D, Wien HC, Singh TP (1975) Some factors affecting soybean viability and emergence in the lowland tropics. IITA grain legume improvement program papers. In: World soybean research conference, Urbana-Champaign, Illinois, United States, pp 1973–1977

Nannipieri P, Muccini L, Ciardi C (1983) Microbial biomass and enzyme activities: production and persistence. Soil Biol Biochem 15(6):679–685

Nannipieri P, Sequi P, Fusi P (1996) Humus and enzyme activity. In: Humic substances in terrestrial ecosystems. Elsevier, pp 293–328

Nannipieri P, Giagnoni L, Landi L, Renella G (2011) Role of phosphatase enzymes in soil. In: Phosphorus in action: biological processes in soil phosphorus cycling, pp 215–243

Naorem A, Jayaraman S, Sinha NK, Mohanty M, Chaudhary RS, Hati KM, Mandal A, Thakur JK, Patra AK, Srinivasarao C, Chaudhari SK, Dalal RC, Lal R (2023) Eight-year impacts of conservation agriculture on soil quality, carbon storage, and carbon emission footprint. Soil Tillage Res 232:105748

Narayana DVV, Sastry G (1985) Soil conservation in India. In: El-Saif SA, Moldenhauer WC, Lo A (eds) Soil erosion and conservation. Soil Conservation Society of America, Ankeny, pp 3–9

Nathan OO, Monica MM, Jayne MN, Isaya S, George N, Daniel MN (2022) Nutrient and organic carbon losses by erosion, and their economic and environmental implications in the drylands of Kenya. Environ Challenge 7:100519

Nayak AK, Sharma DK, Mishra VK, Minhas PS, Verma CL (2008) Reclamation of saline-sodic soil under a rice–wheat system by horizontal surface flushing. Soil Use Manag 24(4):337–343

Nciizah AD, Wakindiki II (2014) Rainfall pattern effects on crusting, infiltration and erodibility in some South African soils with various texture and mineralogy. Water SA 40(1):57–63

Nciizah AD, Wakindiki II (2015a) Physical indicators of soil erosion, aggregate stability and erodibility. Arch Agron Soil Sci 61(6):827–842

Nciizah AD, Wakindiki II (2015b) Soil sealing and crusting effects on infiltration rate: a critical review of shortfalls in prediction models and solutions. Arch Agron Soil Sci 61(9):1211–1230

Nelson PJ (1997) To hold the land: soil erosion, agricultural scientists, and the development of conservation tillage techniques. Agric Hist 71(1):71–90

Niaz S, Wehr BJ, Dalal RC (2022) Organic amendments and gypsum reduce dispersion and increase aggregation of two sodic Vertisols. Geoderma 425:116047

Nimmo JR (2003) Porosity and pore size distribution. In: Hillel D (ed) Encyclopaedia of soils in the environment, vol 3. Elsevier, London, pp 295–303

Nimmo JR (2004) Porosity and pore size distribution. Encyclopedia of Soils in the Environment 3 (1):295–303

Nordstrom KF, Hotta S (2004) Wind erosion from cropland in the USA: a review of problems, solutions, and prospects. Geoderma 121(3–4):157–167

Norwell WA, Lindsay WL (1972) Reactions of DTPA chelates of Fe, Zn, Cu and Mn with soils. Soil Sci Soc Am Proc 36:778–789

Nziguheba G, Smolders E (2008) Inputs of trace elements in agricultural soils via phosphate fertilizers in European countries. Sci Total Environ 390(1):53–57

Oades JM (1984) Soil organic matter and structural stability: mechanisms and implications for management. Plant Soil 76:319–337

Oades JM (1988) The retention of organic matter in soils. Biogeochemistry 5:35–70

Oades JM, Ladd JN (1977) Biochemical properties: carbon and nitrogen metabolism. In: Russell JS, Greacen EL (eds) Soil factors in crop production in a semi-arid environment. St. Lucia, Univ Queenslands Press

Oades JM, Waters AG (1991) Aggregate hierarchy in soils. Aust J Soil Res 29:815–828

Obi AO (1989) Long-term effects of the continuous cultivation of a tropical Ultisol in south-western Nigeria. Expl Agric 25:207–215

Odendo M, Obare G, Salasya B (2010) Farmers' perceptions and knowledge of soil fertility degradation in two contrasting sites in western Kenya. Land Degrad Dev 21(6):557–564

Oldeman LR (1997) Soil degradation: a threat to food security? In: International conference on time ecology: time for soil culturetemporal perspectives on sustainable use of soil, Tutzing, pp 1–16

Oliver MA (1997) Soil and human health: a review. Eur J Soil Sci 48(4):573–592

Onstad CA, Foster GR (1975) Erosion modelling on a watershed. Trans Am Soc Agr Eng 18:288–292

Osman KT (2018) Saline and sodic soils. In: Management of soil problems. Springer, pp 255–298. https://doi.org/10.1007/978-3-319-75527-4_10

Page KL, Allen DE, Dalal RC, Slattery W (2009) Processes and magnitude of CO, CH , and N O fluxes from liming of Australian acidic soils: a review. Aust J Soil Res 47:747–762

Page KL, Bell M, Dalal RC (2013a) Changes in total soil organic carbon stocks and carbon fractions in sugarcane systems as affected by tillage and trash management in Queensland, Australia. Soil Res 51(8):608–614

Page KL, Dalal RC, Pringle MJ, Bell M, Dang YP, Radford B, Bailey K (2013b) Organic carbon stocks in cropping soils of Queensland, Australia, as affected by tillage management, climate, and soil characteristics. Soil Res 1(8):596–607

Page KL, Dalal RC, Reeves SH, Wang WJ, Jayaraman S, Dang YP (2020a) Changes in soil organic carbon and nitrogen after 47 years with different tillage, stubble, and fertiliser management in a Vertisol of north-eastern Australia. Soil Res 58(4):346–355

Page KL, Dang YP, Dalal RC (2020b) The ability of conservation agriculture to conserve soil organic carbon and the subsequent impact on soil physical, chemical, and biological properties and yield. Front Sustain Food Syst 4:31

Pagliai M, Vignozzi N (2002) The soil pore system as an indicator of soil quality. Adv Geo Ecol 35:69–80

Pal R, Poonia SR (1990) Models of solute transport in soils. Indian J Agric Sci 60:503–515

Panagos P, Meusburger K, Ballabio C, Borrelli P, Alewell C (2014) Soil erodibility in Europe: a high-resolution dataset based on LUCAS. Sci Total Environ 479:189–200

Pancholy SK, Rice EL (1973) Soil enzymes in relation to old field succession: amylase, cellulase, invertase, dehydrogenase, and urease. Soil Sci Soc Am J 37(1):47–50

Parton WJ (1996) The CENTURY model. In: Evaluation of soil organic matter models: using existing long-term datasets. Springer, Berlin/Heidelberg, pp 283–291

Parton WJ, Rasmussen PE (1994) Long-term effects of crop management in wheat-fallow: II. CENTURY model simulations. Soil Sci Soc Am J 58(2):530–536

Parton WJ, Schimel DS, Cole CV, Ojima DS (1987) Analysis of factors controlling soil organic matter levels in Great Plains grasslands. Soil Sci Soc Am J 51:1173–1179

Parton WJ, Stewart JWB, Cole CV (1988) Dynamics of C, N, P and S in grassland soils: a model. Biogeochemistry 5:109–131

Parton WJ, McKeown R, Kirchner V, Ojima D (1992) CENTURY Users' Manual. Natural Resource Ecology Laboratory, Colorado State University, Ft Collins

Pasricha NS (2017) Conservation agriculture effects on dynamics of soil C and N under climate change scenario. Adv Agron 145:269–312

Pathak P, Wani SP, Sudi R (2011) Long-term effects of management systems on crop yield and soil physical properties of semiarid tropics of Vertisols. Agric Sci 2(4):435–442

Paul EA, Voroney RP (1980) Nutrient and energy flows through soil microbial biomass. Academic

Pauli FW (1967) Soil fertility: a biodynamical approach. Adam Hilger, London

Pavan MA, Bingham FT, Pratt PF (1982) Toxicity of aluminium to coffee in Ultisols and Oxisols amended with $CaSO_4.2H_2O$. Soil Sci Soc Am J 46:1201–1207

Pearson GA (1960) Tolerance of crops to exchangeable sodium. Agricultural Information Bull No 206, U S Dept Agric -ARS, Washington, DC

Peck AJ (1981) Causes and treatment of secondary salinity. In: Rixon AJ (ed) Proceedings of salinity and water quality symposium, Toowoomba, Queensland, pp 5–11

Pendias AK (2004) Soil-plant transfers of trace element- an environmental issue. Geoderma 122:143–149

Pendias AK, Pendias H (1992) Trace elements in soils and plants, 2nd edn. CRC Press, Boca Raton

Pettit NM, Smith ARJ, Freedman RB, Burns RG (1976) Soil urease: activity, stability and kinetic properties. Soil Biol Biochem 8(6):479–484

Pierce FC, Larson WE, Dowdy RH, Graham WA (1983) Productivity of soils: assessing long-term changes due to erosion. J Soil Water Conserv 38:39–44

Pieri C (1979) Etude de la composition de la solution du sol sableuxcultive du Senegal a l'aide de capteursenceramiqueporeux. L'Agronomie Tropicale 34:9–18

Pieri C (1987) Management of acid tropical soils in Africa. In: Sanchez PA, Stoner ER, Pushparajah E (eds) Management of acid tropical soils for sustainable agriculture. International Board for Soil Research and Management, Bangkok, pp 41–61

Pieri C, Oliver R (1988) Assessment of K losses in tropical cropping systems of francophone Africa and Madagascar. In: International Potash Research Institute (ed) Nutrient balance and needs for potassium. International Potash Research Institute, Berne, pp 73–92

Pillai UP, McGarry D (1999) Structure repair of a compacted Vertisol with wet dry cycles and crops. Soil Sci Soc Am J 63(1):201–210

Poonia P, Parihar S (2020) Biochar mitigates soil salinity: a revies. Bull Environ Pharmacol Life Sci 9:169–174

Post WM, Pastor J, Zinke PJ, Stangenberger AG (1985) Global patterns of soil nitrogen storage. Nature (London) 317:613–616

Power JF, Legg JO (1978) Effect of crop residues on the soil chemical environment and nutrient availability. In: Oschwald WR (ed) Crop residue management systems. Am Soc Agron Madison, Wisconsin, pp 85–100

Powlson DS, Stirling CM, Jat ML, Gerard BG, Palm CA, Sanchez PA, Cassman KG (2014) Limited potential of no-tillage agriculture for climate change mitigation. Nat Clim Change 4:678–683

Prasad R, Power JF (1991) Crop residue management. Adv Soil Sci 15:205–251

Prasad R, Gangaiah B, Aipe KC (1999) Effect of crop residue management in a rice–wheat cropping system on growth and yield of crops and on soil fertility. Exp Agric 35(4):427–435

Prasad R, Kumar M, Varma A (2015) Role of PGPR in soil fertility and plant health. In: Plant-growth-promoting rhizobacteria (PGPR) and medicinal plants, Soil biology, vol 42. Springer, Cham, pp 247–260. https://doi.org/10.1007/978-3-319-13401-7_12

Prebble RE (1987) Effect of cultivation on aggregate stability, microaggregation and organic carbon of Vertisols, Division of Soils, divisional report no 91. CSIRO, Melbourne, p 108

Pringle MJ, Allen DE, Dalal RC, Payne JE, Mayer DG, O'Reagain PO, Marchant BP (2011) Soil carbon stock in the tropical rangelands of Australia: effects of soil type and grazing pressure, and determination of sampling requirement. Geoderma 167–168:261–273

Pringle MJ, Allen DE, Phelps DG, Bray SG, Orton TG, Dalal RC (2014). The effect of pasture utilization rate on stocks of soil organic carbon and total nitrogen in a semi-arid tropical grassland. Agric Ecosys Environ 195: 83–90.

Probert ME, Fergus IF, Bridge BJ, McGarry D, Thompson CH, Russell JS (1987) The properties and management of Vertisols. CAB International, Wallingford

Pushparajah E, Bachik AT (1987) Management of acid tropical soils in south-east Asia. In: Sanchez PA, Stoner ER, Pushparajah E (eds) Management of acid tropical soils for sustainable agriculture. International Board for Soil Research and Management, Bangkok, pp 13–39

Qadir M, Qureshi RH, Ahmad N (1998) Horizontal flushing: a promising ameliorative technology for hard saline-sodic and sodic soils. Soil Tillage Res 45(1–2):119–131

Quirk JP (1986) Soil permeability in relation to sodicity and salinity. Philos Trans R Soc Lond A 316:297–317

Quirk JP, Schofield RK (1955) The effect of electrolyte concentration on soil permeability. J Soil Sci 6:163–178

Raj R, Saharia M, Chakma S (2023) Mapping soil erodibility over India. Catena 230:107271. https://doi.org/10.1016/j.catena.2023.107271

Ramana S, Biswas AK, Rao AS (2009) Phytoremediation of cadmium contaminated soils by marigold and chrysanthemum. Natl Acad Sci Lett 32(11/12):333–336

Ramana S, Biswas AK, Singh AB, Ajay ANK, Subba Rao A (2013) Phytoremediation ability of some floricultural plant species. Indian J Plant Physiol 18(2):187–190

Ramana S, Tripathi AK, Kumar A, Dey P, Saha JK, Patra AK (2021) Evaluation of Furcraeafoetida Haw L for phytoremediation of cadmium contaminated soils. Environ Sci Pollut Res 28:14177–14181

Rao DL, Ghai SK (1985) Urease and dehydrogenase activity of alkali and reclaimed soils. Soil Res 23(4):661–665

Rao KV, Kishore R, Singh R (2009) Mole drainage to enhance soybean production in waterlogged Vertisols. J Agric Eng 46(4):54–58

Rao KV, Bajpai A, Gangwar S, Chourasia L, Soni K (2016) Maximising water productivity of wheat crop by adopting drip irrigation. Res Crops 17(1):163–168

Rashmi I, Meena BP, Rajendiran S, Jayaraman S, Joshy CG, Ali S, Mina BL, Kumar K, Kumar A, Kumawat A, Kala S (2024) Can gypsum and organic amendments achieve sustainability, productivity and maintain soil health under soybean-mustard cropping in sodic soils of western India. Soil Tillage Res 240:106075

Rasool R, Kukal SS, Hira GS (2007) Soil physical fertility and crop performance as affected by long term application of FYM and inorganic fertilizers in rice–wheat system. Soil Tillage Res 96(1–2):64–72

Rayment GE (1988) Cadmium in Queensland's primary industries-1988. In: Simpson J, Curnow W (eds) Cadmium Accumulations in Australian Agriculture: National Symposium Canberra, 1–2 March 1988. Aust. Gov. Publish. Service. Dept. of Prim. Ind. Energy Bur. of Rural Res. Proc. No. 2, Canberra, pp 151–159

Rayment GE (2005) Cadmium in sugar cane and vegetable systems of Northeast Australia. Commun Soil Sci Plant Anal 36(4-6):597–608

Regassa H, Elias E, Tekalign M (2023) The nitrogen fertilizer replacement values of incorporated legumes residue to wheat on Vertisols of the Ethiopian highlands. Heliyon 9(11):e22119

Reid K, Schneider K, McConkey B (2018) Components of phosphorus loss from agricultural landscapes, and how to incorporate them into risk assessment tools. Front Earth Sci 6:135

Renard KG, Foster GR, Weesies GA, Porter JP (1991) RUSLE: revised universal soil loss equation. J Soil Water Conserv 46(1):30–33

Renard KG, Foster GR, Weesies GA, McCool DK, Yoder DC (1996) Predicting soil erosion by water: a guide to conservation planning with the Revised Universal Soil Loss Equation (RUSLE). In: Agriculture handbook, vol 703, p 400

Rengasamy P (2002) Transient salinity and subsoil constraints to dryland farming in Australian sodic soils: an overview. Aust J Exp Agric 42:351–361

Rengasamy P, Greene RSB, Ford GW, Mehanni AH (1984) Identification of dispersive behaviour and the management of red brown earths. J Soil Sci 34:723–732

Rengasamy P, North S, Smith A (2010) Diagnosis and management of sodicity and salinity in soil and water in the Murray Irrigation region, The University of Adelaide, South Australia, viewed

02 April–May 2024. https://www.agric.wa.gov.au/dispersive-and-sodic-soils/managing-dispersive-sodic-soils

Rengasamy P, de Lacerda CF, Gheyi HR (2022) Salinity, sodicity and alkalinity. In: Oliveira TS, Bell RW (eds) Subsoil constraints for crop production. Springer, Cham, pp 83–107. https://doi.org/10.1007/978-3-031-00317-2_4

Reuter DJ, Robinson JB (1986) Plant analysis. Inkata Press, Melbourne

Ribeiro BT, de Lima JM, Curi N, de Oliveira GC, da Silva ÉA, Silva BM (2017) Ultrasonic aggregate breakdown of an Oxisol as affected by cavitation intensity. Commun Soil Sci Plant Anal 48(7):818–824

Richards AE, Dalal RC, Schmidt S (2007) Soil carbon turnover and sequestration in native subtropical tree plantations. Soil Biol Biochem 39:2078–2090

Richards AE, Dalal RC, Schmidt S (2009) Carbon storage in a Ferrosol under subtropical forest, tree plantations, and pasture is linked to soil aggregation. Aust J Soil Res 47:341–350

Ritter WF (1989) Nitrate leaching under irrigation in the United States – review. J Environ Sci Health A 24:349–378

McCown RL, Jones RK, Peake DCI (1980) Short-term benefits of zero-tillage in tropical grain production. Proceedings of the Australian Agronomy Conference, April 1980, Queensland Agricultural College, Lawes, Queensland, Australia. p 220

Rocci KS, Lavallee JM, Stewart CE, Cotrufo MF (2021) Soil organic carbon response to global environmental change depends on its distribution between mineral-associated and particulate organic matter: A meta-analysis. Sci Total Environ 793: 148569

Roose EJ (1977) Application of the universal soil loss equation of Wischmeier and Smith in West Africa. In: Greenland DJ, Lal R (eds) Soil conservation and management in the humid tropics. Wiley, Chichester, pp 177–187

Roose EJ (1986) Runoff and erosion before and after clearing depending on the type of crop in western Africa. In: Lal R, Sanchez PA, Cummings RW Jr (eds) Land clearing and development in the tropics. Balkema, Rotterdam, pp 317–330

Rose CW (1985) Developments in soil erosion and depositional models. Adv Soil Sci 2:1–63

Rose CW, Dalal RC (1988a) Erosion and runoff of nitrogen. In: Wilson JR (ed) Advances in nitrogen cycling in agricultural ecosystems. C.A.B. International, Wallingford, pp 312–326

Rose CW, Dalal RC (1988b) Erosion and runoff of nitrogen. In: Wilson JR (ed) Advances in nitrogen cycling in agricultural ecosystems. CAB International, Wallingford, pp 212–235

Ross SM (1993) Organic matter in tropical soils: current conditions, concerns and prospects for conservation. Progress in physical geography 17(3):265–305

Rowell DL (1994) Soil science methods and applications. Longman Scientific and Technical, Harlow

Rowell DL, Wild A (1985) Causes of soil acidification: a summary. Soil Use Manag 1:32–33

Ru-kun L, Bai-fan J, Kwei LC (1982) Phosphorus management for submerged rice soils. Trans 12th Intl Congr Soil Sci Symp 11:182–191

Russell EW (1973) Soil conditions and plant growth, 10th edn. Longmans, London

Russell JS (1976) Comparative salt tolerance of some tropical and temperate legumes and tropical grasses. Aust J Exp Agric Anim Husb 16:103–109

Russell JS (1984) Management of cracking clay soils in rainfed environments. 307–315 in The properties and utilization of cracking clay soils: proceedings of a Symposium, 24-28 Aug 1981, Armidale, New South Wales, Australia In: McGarity JW, Hoult EH, So H (eds). Reviews in Rural Science no. 5. University of New England, Armidale

Russell JS, Jones PN (1996) Continuous, alternate and double crop systems on a Vertisol in subtropical Australia. Aust J Exp Agric 36(7):823–830

Ryden JC, Rolston DE (1983) The measurement of denitrification. In: Freney JR, Simpson JR (eds) Gaseous loss of nitrogen from plant-soil systems. Developments in plant and soil sciences, vol 9. Springer, Dordrecht. https://doi.org/10.1007/978-94-017-1662-8_4

Sahrawat KL (1983) Nitrogen availability indexes for submerged rice soils. Adv Agron 36:415–451

Sahrawat KL, Wani SP, Pathak P, Rego TJ (2010) Managing natural resources of watersheds in the semi-arid tropics for improved soil and water quality: a review. Agric Water Manag 97(3):375–381
Salinity Management Handbook (2011) Salinity management handbook, 2nd edn. Department of Environment and Resource management, the State of Queensland
Samal SK, Rao KK, Poonia SP, Kumar R, Mishra JS, Prakash V, Mondal S, Dwivedi SK, Bhatt BP, Naik SK, Choubey AK, Kumar V, Malik RK, McDonald A (2017) Evaluation of long-term conservation agriculture and crop intensification in rice-wheat rotation of Indo-Gangetic Plains of South Asia: carbon dynamics and productivity. Eur J Agron 90:198–208
Sanchez PA (1976) Properties and management of acid soils in the tropics. Wiley, New York
Sanchez PA (1987) Management of acid soils in the humid tropics of Latin America. In: Sanchez PA, Stoner ER, Pushparajah E (eds) Management of acid tropical soils for sustainable agriculture. International Board for Soil Research and Management, Bangkok, pp 63–107
Sanchez PA (2002) Soil fertility and hunger in Africa. Science 295:2019–2020. https://doi.org/10.1126/science.1065256
Sanchez PA, Benites JR (1987) Low-input cropping for acid soils of the humid tropics: a transition technology between shifting and continuous cultivation. In: Latham M, Ahn P (eds) Land development and management of acid soils in Africa. International Board for Soil Research and Management, Bangkok, pp 85–106
Sanchez PA, Jama BA (2001) Soil fertility replenishment takes off in East and Southern Africa. In: Integrated plant nutrient management in sub-Saharan Africa: from concept to practice. CABI Publishing, Wallingford, pp 23–45
Sanchez PA, Uehara G (1980) Management considerations for acid soils with high phosphorus fixation capacity. In: The role of phosphorus in agriculture, pp 471–514
Sanchez PA, Stoner ER, Pushpa Rajah E (1987) Management of acid tropical soils for sustainable agriculture. In: Proc IBSRAM No. 2. International Board for Soil Research and Management, Bangkok
Sanchez PA, Shepherd KD, Soule MJ, Place FM, Buresh RJ, Izac AMN, Uzo Mokwunye A, Kwesiga FR, Ndiritu CG, Woomer PL (1997) Soil fertility replenishment in Africa: an investment in natural resource capital. In: Replenishing soil fertility in Africa, vol 51, pp 1–46
Sanderman J, Hengl T, Fiske GJ (2018) Correction for Sanderman et al Soil carbon debt of 12,000 years of human land use. Proc Natl Acad Sci USA 115(7):E1700
Santra P, Moharana PC, Kumar M, Soni ML, Pandey CB, Chaudhari SK, Sikka AK (2017) Crop production and economic loss due to wind erosion in hot arid ecosystem of India. Aeolian Res 28:71–82
Sanyal SK, de Datta SK (1991) Chemistry of phosphorus transformations in soil. Adv Soil Sci 16:1–120
Sarangi SK, Mainuddin M, Maji B (2022) Problems, management, and prospects of acid sulphate soils in the Ganges Delta. Soil Syst 6(4):95
Sarker JR, Singh BP, Cowie AL, Fang Y, Collins D, Badgery W, Dalal RC (2018a) Agricultural management practices impacted carbon and nutrient concentrations in soil aggregates, with minimal influence on aggregate stability and total carbon and nutrient stocks in contrasting soils. Soil Tillage Res 178:209–223
Sarker JR, Singh BP, Dougherty WJ, Fang Y, Badgery W, Hoyle FC, Dalal RC, Cowie AL (2018b) Impact of agricultural management practices on the nutrient supply potential of soil organic matter under long-term farming systems. Soil Tillage Res 175:71–81
Sarmah AK, Pillai-McGarry U, Mc Garry D (1996) Repair of the structure of a compacted Vertisol via wet/dry cycles. Soil Tillage Res 38(1–2):17–33
Schepers JS, Raun W (eds) (2008) Nitrogen in agricultural systems. American Society of Agronomy CSSA-SSSA, Inc. https://doi.org/10.2134/agronmonogr49
Schiettecatte W, Gabriels D, Cornelis WM, Hofman G (2008a) Enrichment of organic carbon in sediment transport by interrill and rill erosion processes. Soil Sci Soc Am J 72(1):50–55

Schiettecatte W, Verbist K, Gabriels D (2008b) Assessment of detachment and sediment transport capacity of runoff by field experiments on a silt loam soil. Earth Surf Process Landf 33(8):1302–1314

Schmitthener AF (1985) Problems and progress in control of Phytophthora root rotof soybean. Plant Dis 69:362–368

Schmitthenner AF, Kroetz ME (1982) Pyroxyfur seed treatment for integrated control of Phytophthora root rot of soybean (Abstract). Phytopath 72:971

Schumacher T, Linfdytom MJ, Mokma DL, Nelson WW (1994) Corn yield-erosion relationship of representative loess and till soils in the North Central United-States. J Soil Water Conserv 49 (2):77–81

Schuman GE, Janzen HH, Herrick JE (2002) Soil carbon dynamics and potential carbon sequestration by rangelands. Environ Pollut 116:391–396

Scott BJ, Ridley AM, Conyers MK (2000) Management of soil acidity in long-term pastures of south-eastern Australia: a review. Aust J Exp Agric 40(8):1173–1198

Serrão EAS, Falesi IC, Veiga JD, Teixeira Neto JF (1979) Productivity of cultivated pastures on low fertility soils in the Amazon of Brazil. In: Pasture production in acid soils of the tropics, pp 195–225

Seubert CE, Sanchez PA, Valverde C (1977) Effects of land clearing methods on soil properties of a Ultisol and crop performance in the Amazon jungle of Peru. Trop Agric Trinidad 54:307–321

Shainberg I (1985) The effect of exchangeable sodium and electrolyte concentration on crust formation. Adv Soil Sci 1:101–122

Shainberg I, Letey J (1984) Response of soils to sodic and saline conditions. Hilgardia 52:1–57

Shainberg I, Sumner ME, Miller WP, Farina MPW, Pavan MA, Fey MV (1989) Use of gypsum on soils: a review. Springer, US, pp 1–111

Sharda VN, Dogra P (2013) Assessment of productivity and monetary losses due to water erosion in rainfed crops across different states of India for prioritization and conservation planning. Agric Res 2:382–392

Sharda VN, Mandai D, Ojasvi PR (2013) Identification of soil erosion risk areas for conservation planning in different states of India. J Environ Biol 34(2):219

Sharma DK, Chaudhari SK (2012) Agronomic research in salt affected soils of India: an overview. Indian J Agron 57(3S):175–185

Sharma RK, Kushwah SS, Sharma SK (2000) Improved surface drainage by changing land configuration in flat land. J Agric Eng 37(4):63–66

Sharpley AN (1980) The enrichment of soil phosphorus in runoff sediments. J Environ Qual 99:521–526

Sharpley AN (1985) The selective erosion of plant nutrients in runoff. Soil Sci Soc Am J 49:1527–1534

Shaw RJ (1995) A unified soil property and sodicity model of salt leaching and water movement. PhD thesis, The University of Queensland, Brisbane

Shaw RJ, Thorburn PJ (1985) Prediction of leaching fraction from soil properties, irrigation water and rainfall. Irrig Sci 6:73–83

Shaw RJ, Yule DF (1978) Assessment of soils for irrigation, Emerald, Queensland, Agric Chem Branch Tech Report No 13. Queensland Department of Primary Industries, Brisbane

Shaw RJ, Hughes KK, Thorburn PJ, Dowling AJ (1987) Principles of landscape, soil and water salinity – processes and management options. Part A. In: Landscape, soil and water salinity, Proceedings of the Brisbane regional salinity workshop, Brisbane, May 1987, Queensland Department of Primary Industries' Conference and Workshop Series QC87003

Shaw R, Brebber L, Ahern C, Weinand M (1994) A review of sodicity and sodic soil behaviour in Queensland. Soil Res 32(2):143–172

Shcherbak L, Millar N, Robertson GP (2014) Global meta-analysis of the nonlinear response of soil nitrous oxide (N O) emissions to fertilizer nitrogen. Proc Natl Acad Sci USA 111:9199–9204

Silva G (2011) Nutrient removal rates in grain crops. Michigan State University Extension Bulletin 18:2014

Silva ÉA, Oliveira GC, Silva BM, Carducci CE, Avanzi JC, Serafim ME (2014) Aggregate stability by the high energy moisture characteristic method in an oxisol under differentiated management. Revista Brasileira de Ciência do Solo 38:1633–1642

Silveira ML, Xu S, Adewopo J, Franzluebbers AJ, Buonadio G (2014) Grazing land intensification effects on soil C dynamics in aggregate size fractions of a Spodosol. Geoderma 230:185–193

Silver WL, Ostertag R, Lugo AE (2000) The potential for carbon sequestration through reforestation of abandoned tropical agricultural and pasture lands. Restor Ecol 8:394–407

Silver WL, Perez T, Mayer A, Jones AR (2021) The role of soil in the contribution of food and feed. Philos Trans R Soc B 376(1834):20200181

Singh VK, Dwivedi BS (2006) Yield and nitrogen use efficiency in wheat, and soil fertility status as influenced by substitution of rice with pigeon pea in a rice-wheat cropping system. Aust J Exp Agric 46:1185–1194

Singh M, Wanjari RH, Dwivedi AK, Dalal RC (2012) Yield response to applied nutrients and estimates of N fixation In 33-yearold soybean–wheat experiment on a vertisol. Exp Agric 48:311–325

Singh M, Wanjari RH, Jatav RC (2017) Phosphorus and potassium management under long-term manuring and fertilisation. Indian J Fertil 13(4):98–109

Singh RK, Chaudhary RS, Somasundaram J et al (2020) Soil and nutrients losses under different crop covers in Vertisols of Central India. J. Soils Sediments 20:609–620

Singh M, Wanjari RH, Kumar U (2021) Long-term fertilizer experiments in India: achievements and issues for future research Soil science: fundamentals to recent advances, pp 781–800

Siwik-Ziomek A, Lemanowicz J, Koper J (2016) Arylsulphatase activity and sulphate content in relation to crop rotation and fertilization of soil. Int Agrophys 30(3):359–367

Six J, Jastrow JD (2002) Organic matter turnover. In, Encyclopedia of soil science, p 10

Six J, Paustian K (2014) Aggregate-associated soil organic matter as an ecosystem property and a measurement tool. Soil Biol Biochem 68:A4–A9

Six JA, Elliott ET, Paustian K (2000a) Soil macroaggregate turnover and microaggregate formation: a mechanism for C sequestration under no-tillage agriculture. Soil Biol Biochem 32 (14):2099–2103. https://doi.org/10.1016/S0038-0717(00)00179-6

Six J, Paustian K, Elliott ET, Combrick C (2000b) Soil structure and organic matter: I. Distribution of aggregate-size classes and aggregate-associated carbon. Soil Sci Soc Am J 64:681–689

Six J, Feller C, Denef K, Ogle S, de Moraes Sa JC, Albrecht A (2002) Soil organic matter, biota, and aggregation in temperate and tropical soils – effects of no-tillage. Agronomie 22(7–8):755–775

Six J, Bossuyt H, Degryze S, Denef K (2004) A history of research on the link between (micro) aggregates, soil biota, and soil organic matter dynamics. Soil Tilliage Res 79(1):7–31

Skjemstad JO, Dalal RC (1987) Spectroscopic and chemical differences in organic matter of two vertisols subjected to long periods of cultivation. Aust J Soil Res 25:323–335

Skjemstad JO, Dalal RC, Barron PF (1986) Spectroscopic investigations of cultivations effects on organic matter of Vertisols. Soil Sci Soc Am J 50:354–359

Skjemstad JO, Le Feuvre RP, Prebble RE (1990) Turnover of soil organic matter under pasture as determined by C natural abundance. Aust J Soil Res 28:267–277

Skoglund LG, Brown WM (1988) Effects of tillage regimes and herbicides on Fusarium species associated with corn stalk rot. Can J Plant Path 10(4):332–338

Skujins J (1973) Dehydrogenase an indicator of biological activities in arid soils. Bull Ecol Res Commun 17:235–241

Skujins J (1978) History of abiontic soil enzyme research. In: Burnes RG (ed) Soil enzymes. Academic, London, pp 1–49

Smeck NE (1985) Phosphorus dynamics in soils and landscapes. Geoderma 36(3–4):185–199

Smil V (1999) Nitrogen in crop production: an account of global flows. Global Biogeochem Cycles 13(2):647–662

Smith AE (1982) Herbicides and the soil environment in Canada. Can J Soil Sci 62:433–460

Smith JL, Doran JW (1996) Measurement and use of pH and electrical conductivity for soil quality analysis. In: Methods for assessing soil quality, Special Publication 49. Soil Science Society of America, pp 169–185

Smith JA, Novak JT (1987) Biodegradation of chlorinated phenols in subsurface soils. Water Air Soil Pollut 33:29–42

Smith GD, Yule DF, Coughlan KJ (1984) Soil physical factors in crop production on Vertisols in Queensland, Australia. In: Craswell ET, Isbell RF (eds) Proceedings international workshop on soils. Australian Centre for International Agricultural Research, Townsville, pp 87–104

Smith P, Keesstra SD, Silver WL, Adhya TK, de Deyn GB, Carvalheiro LG, Giltrap DL, Renfprth P, Cheng K, Sarkar B, Saco PM, Scow K, Smith J, Morel JC, Thiele-Bruhn LR, McElwee P (2021) Soil-derived nature's contributions to people and their contribution to the UN sustainable development goals. Philos Trans R Soc B 376(1834):20200169

Smith MR, Dinglasan E, Veneklaas E, Polania J, Rao IM, Beebe SE, Merchant A (2022) Effect of drought and low P on yield and nutritional content in common bean. Front Plant Sci 13:814325

SMSS Technical Monograph, Soil Management Support Services, USA, 1988, 18, 165–181 ref. 45

Soler-Rovira P, Soler-Soler J, Soler-Rovira J, Polo A (1995) Agricultural use of sewage sludge and its regulation. Fertil Res 43:173–177

Somasundaram J (2001) Evaluation of sewage sludge-coir pith pellets on fodder crops and biotransfer of heavy metals, Doctoral dissertation, Tamil Nadu Agricultural University, Coimbatore

Somasundaram J, Reeves S, Wang W, Heenan M, Dalal RC (2017) Impact of 47 years of no tillage and stubble retention on soil aggregation and carbon distribution in a Vertisol. Land Degrad Dev 28:1589–1602

Somasundaram J, Sinha NK, Dalal RC, Lal R, Mohanty M, Naorem AK, Hati KM, Chaudhary RS, Biswas AK, Patra AK, Chaudhari SK (2020) No-till farming and conservation agriculture in South Asia – issues, challenges, prospects, and benefits. Crit Rev Plant Sci 39:236–279

Sorenson LH (1981) Carbon-nitrogen relationships during the humification of cellulose in soils containing different amounts of clay. Soil Biol Biochem 13:313–321

Sparling GP (1992) Ratio of microbial biomass carbon to soil organic carbon as a sensitive indicator of changes in soil organic matter. Aust J Soil Res 30(2):195–207

Speir TW (1977) Studies on a climosequence of soils in tussock grasslands II. Urease, phosphatase, and sulphatase activities of topsoils and their relationships with other properties including plant available sulphur. New Zealand J Sci 20:159–166

Speir TW, Ross DJ (1978) Soil phosphatase and sulphatase. In: Burns RG (ed) Soil enzymes. Academic, London, pp 197–250

Spoor G (2006) Alleviation of soil compaction: requirements equipment and techniques. Soil Use Manag 22:113–122

Srinivasarao CH, Vittal KP, Tiwari KN, Gajbhiye PN, Kundu S (2007) Categorisation of soils based on potassium reserves and production systems: implications in K management. Soil Res 45(6):438–447

Srinivasarao C, Lal R, Kundu S, Babu MBBP, Venkateswarlu B, Singh AK (2014) Soil carbon sequestration in rainfed production systems in the semiarid tropics of India. Sci Total Environ 487:587–603

Srinivasarao C, Kundu S, Yashavanth BS, Rakesh S, Akbari KN, Sutaria GS, Vora VD, Hirpara DS, Gopinath KA, Chary GR, Prasad JV (2021) Influence of 16 years of fertilization and manuring on carbon sequestration and agronomic productivity of groundnut in vertisol of semi-arid tropics of Western India. Carbon Manage 12(1):13–24

Srivastava SC, Singh JS (1989) Effect of cultivation on microbial carbon and nitrogen in dry tropical forest soil. Biol Fertil Soils 8:343–348

Standley J, Hunter HM, Thomas GA, Blight GW, Webb AA (1990) Tillage and crop residue management affect Vertisol properties and grain sorghum growth over seven years in the semi-arid subtropics. II. Changes in soil properties. Soil Tillage Res 18:367–388

Stanford G (1982) Assessment of soil nitrogen availability. In: Stevenson FJ (ed) Nitrogen in agricultural soils. American Society of Agronomy, Madison, pp 651–688

Steffen W, Richardson K, Rockstrom J, Cornell SE, Fetzer I, Bennett EM, Biggs R, Carpenter SR, De Vries W, De Wit CA, Folke C (2015) Planetary boundaries: guiding human development on a changing planet. Science 347(6223):1259855

Stepniewski W, Gliński J, Ball BC (1994) Effects of compaction on soil aeration properties. In: Developments in agricultural engineering series, vol 11. Elsevier, pp 167–189

Stevens GT, Gladbach P, Motavalli DD (2005) Soil calcium: magnesium ratios and lime recommendations for cotton. J Cotton Sci 9:65–71. https://www.cotton.org/journal/2005-09/2/65.cfm

Stevenson FJ (1982) Humus chemistry. Wiley, New York

Stevenson FJ (1986) Cycles of soil. Wiley, New York

Stoate C, Baldi A, Beja P, Boatman ND, Herzon I, van Doorn A, de Snoo GR, Rakosy L, Ramwell C (2009) Ecological impacts of early 21st century agricultural change in Europe – a review. J Environ Manag 91:22–46

Stocking M (1981) A working model for the estimation of soil loss suitable for underdeveloped areas, Occasional Paper 15. University of East Anglia, Norwich

Stocking M, Peake L (1986) Crop yield losses from the erosion of Alfisols. Trop Agric Trinidad 63:41–45

Stomp TJ, Dordas C, Baranger A, de Rijk J, Dong B, Evers J, Gu C, Li L, Simon J, Jensen ES, Wang Q, Wang Y, Wang Z, Xu H, Zhang C, Zhang L, Zhang W, Bedoussac L, van der Werf W (2020) Designing intercrops for high yield, yield stability and efficient use of resources: are there principles? Adv Agron 160:1–50

Stout JD, Goh KM, Rafter TA (1981) Chemistry and turnover of naturally occurring resistant organic compounds in soil. In: Paul EA, Ladd JN (eds) Soil biochemistry, vol 5. Marcel Dekker, New York, pp 1–73

Strickling E (1950) The effect of soybeans on volume weight and water stability of soil aggregates, soil organic matter and crop yield. Soil Sci Soc Am Proc 15:30–33

Suhail FM, Fahmi AH (2020) Biodegradation of pesticides in soil amended with different organic matters. Indian J Ecol 7(4):1003–1008

Sun W, Canadell JG, Yu L, Yu L, Zhang W, Smith P, Fischer T, Huang Y (2020) Climate drives global soil carbon sequestration and crop yield changes under conservation agriculture. Glob Change Biol 26(6): 3325–3335.

Sumner ME, Noble AD (2003) Soil acidification: the world story. In: Handbook of soil acidity. CRC Press, pp 15–42

Sundermeier AP, Islam KR, Raut Y, Reeder RC, Dick WA (2011) Continuous no-till impacts on soil biophysical carbon sequestration. Soil Sci Soc Am J 75:1779–1788

Suriyagoda LD, Dissanayake O, Kodithuwakku V, Maduwanthi I, Dissanayaka N, Chandrajith R (2022) Accumulation of essential mineral and toxic trace elements in crops and soils of vegetable cropping systems in central highlands of Sri Lanka. J Agric Sci 160(1–2):86–97

Swift MJ, Cook AG, Perfect TJ (1980) The effects of changing agricultural practice on the biology of a forest soil in the subhumid tropics II. Decomposition. In: Furtado JI (ed) Tropical ecology and development Part I. International Society for Tropical Ecology, Kuala Lumpur, pp 541–548

Szabolcs I (1979) Review of research on salt-affected soils. Nat Resour Res 15:137. UNESCO, Paris

Tabatabai MA, Bremner JM (1970) Factors affecting soil arylsulfatase activity. Soil Sci Soc Am J 34(3):427–429

Takala B (2019) Soil acidity and its management options in western Ethiopia. J Environ Earth Sci 9 (10):2224–3216

Tamene L, Adimassu Z, Aynekulu E, Yaekob T (2017) Estimating landscape susceptibility to soil erosion using a GIS-based approach in Northern Ethiopia. Intl Soil Water Cons Res 5(3): pp. 221–230

Tamene L, Adimassu Z, Ellison J, Yaekob T, Woldearegay K, Mekonnen K, Thorne P, Le QB (2017) Mapping soil erosion hotspots and assessing the potential impacts of land management practices in the highlands of Ethiopia. Geomorphology 292: 153–163

Tan KH (2010) Principles of soil chemistry. CRC Press, pp 1–390. https://doi.org/10.1201/9781439894606

Tan ZX, Lal R, Wiebe KD (2005) Global soil nutrient depletion and yield reduction. J Sustain Agric 26:123–146. https://doi.org/10.1300/J064v26n0110

Tan Z, Lal R, Owens L, Izaurralde RC (2007) Distribution of light and heavy fractions of soil organic carbon as related to land use and tillage practice. Soil Tillage Res 92(1–2):53–59

Tandon HLS, Rego TJ (1988) Management of fertiliser nutrients other than nitrogen in the semi-arid tropics of India. In: Christianson CB (ed) Soil fertility and fertilizer management in semi-arid tropical India. International Fertilizer Development Centre, Muscle Shoals, pp 71–88

Tang S, Guo J, Li S, Li J, Xie S, Zhai X, Wang C, Zhang Y, Wang K (2019) Synthesis of soil carbon losses in response to conversion of grassland to agriculture land. Soil Tillage Res 185:29–35

Taylor HM (1980) Mechanical impedance to root growth. In: Priorities for alleviating soil-related constraints to food production in the tropics. IRRI, Los Baños, pp 389–404

Taylor HM, Klepper B (1974) Water relations of cotton: I. Root growth and water use as related to top growth and soil water content. Agron J 66:584–588

Thakuria D, Hazarika S, Krishnappa R (2016) Soil acidity and management options. Indian J Fertil 12(12):40–56

Theng BKG (1987) Clay-humic interactions and soil aggregate stability. In: Rengasamy P (ed) Soil structure and aggregate stability. Vic Dep Agric Rural Affairs, Melbourne, pp 32–73

Thomas GA, Standley J, Hunter HM, Blight GW, Webb AA (1990) Tillage and crop residue management affect Vertisol properties and grain sorghum growth over seven years in the semi-arid subtropics. 3. Crop growth, water use and nutrient balance. Soil Tillage Res 18:367–388

Thomaz EL, Araujo-Junior CF, Vendrame PR, de Melo TR (2022) Mechanisms of aggregate breakdown in (sub) tropical soils: effects of the hierarchical resistance. Catena 216:106377

Thompson JP (1987) Decline of Vesicular-arbuscular mycorrhizae in long fallow disorder of field corps and its expression in phosphorus deficiency in sunflower. Aust J Agric Res 38:847–867

Thompson JP, Mackenzie J, Amos R (1995) Root-lesion nematode (Pratylenchus thornei) limits response of wheat but not barley to stored soil moisture in the Hermitage long-term tillage experiment. Aust J Exp Agric 35(7):1049–1055

Tiessen H, Stewart JWB (1983) Particle size fractions and their use in studies of soil organic matter. II. Cultivation effects on organic matter composition in size fractions. Soil Sci Soc Am J 47:509–514

Tiessen H, Stewart JWB, Moir JO (1983) Changes in organic and inorganic phosphorus composition of two grassland soils and their particle size fractions during 60–90 years of cultivation. J Soil Sci 34:815–823

Tiessen H, Cuevas E, Chacon P (1994) The role of soil organic matter in sustaining soil fertility. Nature 71(6500):783–785

Tinker PB, Ingram JS, Struwe S (1996) Effects of slash-and-burn agriculture and deforestation on climate change. Agric Ecosyst Environ 58(1):13–22

Tisdale SL, Nelson NL (1975) Soil fertility and fertilizers. Mac Millan Co. Inc., New York

Tisdall JM (1994) Possible role of soil microorganisms in aggregation in soils. Plant Soil 159:115–121

Tisdall JM, Oades JM (1980) The effect of crop rotation on aggregation in a red-brown earth. Aust J Soil Res 18:423–433

Tisdall JM, Oades JM (1982) Organic matter and water-stable aggregates in soils. J Soil Sci 33:141–163

Trimble JL, van Hook RI (1984) Biomass for energy: the environmental issues. Biomass 6:3–13

Troeh FR, Thompson LM (2005a) Soil physical properties. In: Soils and soil fertility. Blackwell, Iowa, pp 37–63

Troeh FR, Thompson LM (2005b) Soils and soil fertility, vol 489. Blackwell, Iowa, pp 1–504

Tullberg JN, Yule DF, McGarry D (2007) Controlled traffic farming – from research to adoption in Australia. Soil Tillage Res 97:272–281
UNEP (1982) World soil policy. UN Environment Programme, Nairobi
United Nations (2015) Sustainable development goals: the 17 goals. UN Department of Economic and Social Affairs, Sustainable Development Goals. Available at www.https://sdgs.un.org/goals
USDA-Agricultural Research Service (1961) A universal equation for measuring wind erosion. USDA-ARS.22-69, 22 pp illus USDA-NRCS (2011) Soil quality indicators, pp 1–2
USSL (1954) Diagnosis and improvement of saline and alkali soils. US Department of Agriculture No 60, Washington, DC
Utobo EB, Tewari L (2015) Soil enzymes as bioindicators of soil ecosystem status. Appl Ecol Environ Res 13(1):147–169
Utuk IO, Daniel EE (2015) Land degradation: a threat to food security: a global assessment. J Environ Earth Sci 5(8): 13–21
Vachaud G, Vauclin M, Imbernon J, Dancette C (1982) Large scale determination of flux of water and solute through the soil in situ, in the semi-arid zone taking into account spatial variability. Trans 12th Int Congr Soil Sci 6:13
Valentin C, Bresson LM (2020) Soil crusting. In: Methods for assessment of soil degradation. CRC Press, pp 89–107
Van Beek CGEM, Van Breemen N (1973) The alkalinity of alkali soils. J Soil Sci 24(1):129–136
Van Breeman N, Driscoll CT, Mulder J (1984) Acidic deposition and internal proton sources in acidification of soils and waters. Nature (London) 307:599–604
Van Breemen N (1975) Acidification and deacidification of coastal plain soils as a result of periodic flooding. Soil Sci Soc Am J 39(6):1153–1157
Van Breemen N (1976) Genesis and solution chemistry of acid sulfate soils in Thailand. Wageningen University and Research
Van Breemen NV (1982) Genesis, morphology, and classification of acid sulfate soils in coastal plains. Acid Sulfate Weathering 10:95–108
Van der Linden AMA, van Veen JA, Frissel MJ (1987) Modelling soil organic matter levels after long-term applications of crop residues and farmyard and green manures. Plant Soil 101:21–28
Van Lancker J, Bruggeman HY (1982) Rain erosion in the Kizingwe watershed. Facultes de Sciences Agronomiques, Rijks-universiteit Gent, Bujumbura
Van Veen JA, Kuikman PJ (1990) Soil structural aspects of decomposition of organic matter by microorganisms. Biogeochemistry 1:213–233
Van Veen JA, Paul EA (1981) Organic C dynamics in grassland soils I. Background information and computer simulation. Can J Soil Sci 61:185–201
Van Wambeke A, Eswaran H, Herbillon AJ, Comerma J (1983) Oxisols. In: Developments in soil science, vol 11. Elsevier, pp 325–354
Velmourougane K, Venugopalan MV, Bhattacharyya T et al (2013) Urease activity in various agro-ecological sub-regions of black soil regions of India. Proc Natl Acad Sci India Sect B Biol Sci 83:513–524. https://doi.org/10.1007/s40011-013-0162-1
Veloso MG, Cecagno D, Bayer C (2019) Legume cover crops under no-tillage favour organo-mineral association in microaggregates and soil C accumulation. Soil Tillage Res 190:139–146
Verberne ELJ, Hassink J, De Willigen P, Groot JJR, Van Veen JA (1990) Modelling organic matter dynamics in different soils. Neth J Agric Sci 38:221–238
Verbree DA, Duiker SW, Kleinman PJ (2010) Runoff losses of sediment and phosphorus from no-till and cultivated soils receiving dairy manure. J Environ Qual 39(5):1762–1770
Vessey JK (2003) Plant growth promoting rhizobacteria as biofertilizers. Plant Soil 255:571–586
Virto I, Barré P, Chenu C (2008) Microaggregation and organic matter storage at the silt-size scale. Geoderma 46(1–2):326–335
Vitorello VA, Cerri CC, Andreux F, Feller C, Victoria RL (1989) Organic matter and natural carbon-13 distribution in forested and cultivated oxisols. Soil Sci Soc Am J 53:773–778
Vitosh ML, Johnson JW, Mengel DB (1995a) Nitrogen losses from soil. In: Tri-state fertilizer recommendations for corn, soybeans, wheat, and alfalfa, p 24

Vitosh ML, Johnson JW, Mengel DB (1995b) TH-state fertilizer recommendations for corn, soybeans, wheat, and alfalfa Ext Bull- 567. Michigan State University, East Lansing

Vitousek PM, Matson PA (1988) Nitrogen transformations in a range of tropical forest soils. Soil Biol Biochem 20:361–367

Vitousek PM et al (2009) Nutrient imbalances in agricultural development. Science 324:1519–1520. https://doi.org/10.1126/science.1170261

Vityakon P (2005) Soil organic matter loss and fertility degradation under different agricultural land uses in sandy soils of Northeast Thailand and the use of organic materials of different qualities as a possible restoration measure. In: Management of tropical sandy soils for sustainable agriculture, p 533

Voroney RP, Paul EA (1984) Determination of kc and kn in situ for calibration of the chloroform fumigation incubation method. Soil Biol Biochem 16:9–14

Vukicevich E, Lowery T, Bowen P, Úrbez-Torres JR, Hart M (2016) Cover crops to increase soil microbial diversity and mitigate decline in perennial agriculture: a review. Agron Sustain Dev 36:1–4

Vy N, Thi NT (1989) Management of sloping soils for food production in Vietnam. In: Craswell ET, Pushparajah E (eds) Management of acid soils in the humid tropics of Asia, ACIAR monograph No 13. Australian Centre for International Agricultural Research, Canberra, pp 110–114

Wagner LE (2013) A history of wind erosion prediction models in the United States Department of Agriculture: The Wind Erosion Prediction System (WEPS). Aeolian Res 10:9–24

Walker J, Raison RJ, Khanna PK (1986) Fire. In: Russell JS, Isbell RF (eds) Australian soils, the human impact, Australian Society of Soil Science. University of Queensland Press, St Lucia, pp 186–216

Wallace A (1966) Current topics in plant nutrition. Edwards Bros, Ann Arbor

Wallenstein M (2017) To restore our soils, feed the microbes. The Conversation 27

Wang W, Dalal RC (2015) Nitrogen management is the key for low-emission wheat production in Australia: a life cycle perspective. Eur J Agron 66:74–82

Wang Y, Amundson R, Trumbore S (1999) The impact of land use change on C turnover in soils. Global Biogeochem Cycles 3(1):47–57

Wang WJ, Dalal RC, Moody PW (2004) Soil carbon sequestration and density distribution in a Vertosol under different farming practices. Soil Res 42(8):875–882

Wang W, Dalal RC, Reeves SH, Butterbach-Bahl KL, Kiese R (2011) Greenhouse gas fluxes from an Australian subtropical cropland under long-term contrasting management regimes. Glob Change Biol 17(10):3089–3101

Wang ZG, Jin X, Bao XG, Li XF, Zhao JH, Sun JH, Christie P, Li L (2014) Intercropping enhances productivity and maintains the most soil fertility properties relative to sole cropping. PLoS One 9(12):e113984

Wang WJ, Reeves SH, Salter B, Moody PW, Dalal RC (2016) Effects of urea formulations, application rates, and crop residue retention on NO emissions from sugarcane fields in Australia. Agric Ecosyst Environ 216:137–146

Wang L, Huang X, Fang NF, Niu YH, Wang TW, Shi ZH (2019a) Selective transport of soil organic and inorganic carbon in eroded sediment in response to raindrop sizes and inflow rates in rainstorms. J Hydrol 575:42–53

Wang Y, Ying H, Yin Y, Zheng H, Cui Z (2019b) Estimating soil nitrate leaching of nitrogen fertilizer from global meta-analysis. Sci Total Environ 20(657):96–102

Wang X, Sial MU, Bashir MA, Bilal M, Raza QU, Ali Raza HM, Rehim A, Geng Y (2022) Pesticides xenobiotics in soil ecosystem and their remediation approaches. Sustain For 14:3353

Wang C, Xie Y, Tan Z (2024) Soil potassium depletion in global cereal croplands and its implications. Sci Total Environ 07:167875

Wani SP, Srinivasarao C, Rego TJ, Pardhasaradhi G, Roy S (2007) On-farm nutrient depletion and buildup in vertisols under soybean (Glycine max) based cropping systems in semi-arid Central India. Indian J Dryland Agric Res Dev 22(1):69–73

Ward NJ, Sullivan LA, Fyfe DM (2004) The process of sulfide oxidation in some acid sulfate soil materials. Soil Res 42(4):449–458

Waring SA, Teakle LJH (1960) Fertility investigations on the black earth wheat lands of the Darling Downs, Queensland. III. Mineral nitrogen in the soil and its relation to the wheat crop. Aust J Agric Res 11:27–41

Wei X, Shao M, Gale W, Li L (2014) Global pattern of soil carbon losses due to the conversion of forests to agricultural land. Sci Rep 4(1):4062

Wei Y, Wu X, Cai C, Wang J, Xia J, Wang J, Guo Z, Yuan Z (2019) Impact of erosion-induced land degradation on rainfall infiltration in different types of soils under field simulation. Land Degrad Dev 30(14):1751–1764

Weiskopf SR, Isbell F, Arce-Plata MI et al (2024) Biodiversity loss reduces global terrestrial carbon storage. Nat Commun 15:4354. https://doi.org/10.1038/s41467-024-47872-7

Wells HD, Bell DK, Jaworski CA (1972) Efficacy of Trichoderma harzianum as a biocontrol for *Sclerotium rolfsii*. Phytopathology 62:442–447

Wen QX, Lin XX (1986) On the content and characteristics of soil organic matter in south China. In: Li CK et al (eds) Proc Intern Symp Red Soils. Elsevier Science, Amsterdam, pp 309–329

West TO, Marland G (2002) Net carbon flux from agricultural ecosystems: methodology for full carbon cycle analyses. Environ Pollut 116(3):439–444

West TO, Post WM (2002) Soil organic carbon sequestration rates by tillage and crop rotation: a global data analysis. Soil Sci Soc Am J 66(6):1930–1946

White RE (1988) Leaching. In: Wilson JR (ed) Advances in nitrogen cycling in agricultural ecosystems. C.A.B. International, Wallingford, pp 193–211

White PJ, Foley SG, Mayne A (1988) Effect of soil erosion on crop production, Queensland Wheat Research Institute Biennial Report 1986–88, pp 79–80

Wielemaker WG (1984) Soil formation by termites. A study in the Kisii area, Kenya. Doctoral Thesis, Agr Univ, Wageningen

Wiesmeier M, Poeplau C, Sierra CA, Maier H, Frühauf C, Hübner R, Kühnel A, Spörlein P, Geuß U, Hangen E, Schilling B (2016) Projected loss of soil organic carbon in temperate agricultural soils in the 21st century: effects of climate change and carbon input trends. Scientific Rep 6 (1):32525

Wiesmeier M, Urbanski L, Hobley E, Lang B, von Lützow M, Marin-Spiotta E, van Wesemael B, Rabot E, Ließ M, Garcia-Franco N, Wollschläger U (2019) Soil organic carbon storage as a key function of soils-a review of drivers and indicators at various scales. Geoderma 333:149–162

Wildermuth H (1993) Populationsbiologie von Leucorrhinia pectoralis (Charpentier)(Anisoptera: Libellulidae). Libellula 12(3/4):269–275

Wildermuth GB, Warcup JH, Rovira AD (1984) Growth of Gaeumannomyces graminis var. tritici in soil in the presence and absence of wheat roots. Trans British Myco Soc 82(3):435–441

Wildermuth GB, Tinline RD, McNamara RB (1992) Assessment of yield loss caused by common root rot in wheat cultivars in Queensland. Aus J Agric Res 43:43–45

Wildermuth GB, Robertson LN, Thompson JP (1994) Soil biological changes affect sustainability. In: Clarke AL, Wylie PJ (eds) Sustainable agriculture in sub-tropical Australia. Department of Primary Industries, Brisbane, pp 192–220

Wildermuth GB, Thompson JP, Robertson LN (1997) Biological changes: diseases, insects and beneficial organisms. In: Clarke AL, Wylie PB (eds) Sustainable crop production in the sub-tropics. Queensland Department of Primary Industries, QI 97035, Brisbane, pp 112–130

Willett IR, Higgins ML (1978) Phosphate sorption by reduced and reoxidized rice soils. Soil Res 16 (3):319–326

Williams CH (1980) Soil acidification under clover pasture. Aust J Exp Agric Anim Husb 20:561–567

Williams CH, Lipsett J (1961) Fertility changes in soils cultivated for wheat in southern New South Wales. Aust J Agric Res 2:612–620

Williams J, Probert ME (1984) Characterization of the soil-climate constraints for predicting pasture production in the semi-arid tropics. Proceedings of International Workshop on Soils, September 1983, Townsville. ACIAR, Canberra, pp 61–75

Williams JR, Renard KG (1985) Assessment of soil erosion and crop productivity with process models, EPIC. In: Follett RF, Stewart BA (eds) Soil erosion and crop productivity. American Society of Agronomy, Madison, pp 68–104

Williams JR, Allmaras RR, Renard KG, Lyles L, Moldenhauer WC, Langdale GW, Meyer LD, Rawls WJ, Darby G, Daniels R, Magleby R (1981) Soil erosion effects on soil productivity: a research perspective. J Soil Water Conserv 38:381–383

Winkler K, Fichs R, Rounsevell M, Herold M (2021) Global land use changes are four times greater than previously estimated. Nat Commun 12:2501

Wischmeier WH (1984) The USLE: some reflections. J Soil Water Conserv 39(2):105–107

Wischmeier WH, Smith DD (1978) Predicting rainfall erosion losses – a guide to conservation planning, Agriculture Handbook No 537. U. S. Department of Agriculture, Washington, DC

Witt GB, Noël MV, Bird MI, Beeton RJS, Menzies NW (2011) Carbon sequestration and biodiversity restoration potential of semi-arid mulga lands of Australia interpreted from long-term grazing exclosures. Agric Ecosyst Environ 141:108–118

WMO (1983) Meteorological aspects of certain processes affecting soil degradation – especially erosion. World Meteorological Organisation No.178, Rome

Wong MTF, Wild A, Juo ASR (1987) Retarded leaching of nitrate measured in monolith lysimeters in south-east Nigeria. J Soil Sci 38:511–518

Woodruff CM (1949) Estimating the nitrogen delivery of soil from the soil organic matter determination as reflected by Sanbourn field. Soil Sci Soc Proc 14:208–212

Woodruff NP, Siddoway FH (1965) A wind erosion equation. Soil Sci Soc Am Proc 29:602–608

WRB, World reference base for soil resources (2014) Food and Agriculture Organization of the United Nations Rome, pp 1–303

Wu W, Ma B (2015) Integrated nutrient management (INM) for sustaining crop productivity and reducing environmental impact: a review. Sci Total Environ 512:415–427

Xiao S, Zhang W, Ye Y, Zhao J, Wang K (2017) Soil aggregate mediates the impacts of land uses on organic carbon, total nitrogen, and microbial activity in a Karst ecosystem. Sci Rep 7 (1):41402

Yan M, Zhang X, Liu K, Lou Y, Wang Y (2022) Particle size primarily shifts chemical composition of organic matter under longterm fertilization in paddy soil. Eur J Soil Sci 73(1):e13170

Yang G, Sun R, Jing Y, Xiong M, Li J, Chen L (2022) Global assessment of wind erosion based on a spatially distributed RWEQ model. Prog Phys Geogr Earth Environ 46(1):28–42

Yin WA, Tao LI, Li XK, Tao RE, Cong RH, Lu JW (2015) Nutrient deficiency limits population development, yield formation, and nutrient uptake of direct sown winter oilseed rape. J Integra Agric 14(4):670–680

Yoder D, Lown J (1995) The future of RUSLE: inside the new revised universal soil loss equation. J Soil Water Conserv 50(5):484–489

Young A (1984) Land evaluation for agro forestry: the tasks ahead. ICRAF Working Paper No. 24, Nairobi

Young A (1989) Agroforestry for soil erosion. C.A.B. International, Wallingford

Young RA, Mutchler CK (1977) Erodibility of some Minnesota soils. J Soil Water Conserv 32:180–182

Young JL, Spycher G (1979) Water dispersible soil organic mineral particles. I. Carbon and nitrogen distribution. Soil Sci Soc Am J 43:324–328

Yuan Y, Liu Q, Zheng H, Li M, Liu Y, Wang X, Peng Y, Luo X, Li F, Li X, Xing B (2023) Biochar as a sustainable tool for improving the health of salt-affected soils. Soil Environ Health 1:100033

Zahedifard N, Shahbazi K, Mohammadi MH, Golchin A, Moshiri F, Martinez CE (2024) Soil organic carbon fractions in cultivated calcareous soils. Eur Soil Sci 20:1–4

Zanini E, Bonifacio E, Albertson JD, Nielsen DR (1998) Topsoil aggregate breakdown under water-saturated conditions. Soil Sci 163(4):288–298

Zantua MI, Bremner JM (1977) Stability of urease in soils. Soil Biol Biochem 9:135–140

Zhang Y, Bhattacharyya R, Dalal RC, Wang P, Menzies NW, Kopittke PM (2019) Impact of land use change and soil type on total phosphorus and its fractions in soil aggregates. Land Degrad Dev 31:828–841

Zhang Y, Finn D, Bhattacharya R, Dennis PG, Doolette AL, Smernik RJ, Dalal RC, Meyer G, Lombi E, Klysubun W, Jones AR, Wang P, Menzies NW, Kopittke PM (2021) Long-term changes in land use influence phosphorus concentrations, speciation, and cycling within sub-tropical soils. Geoderma 393:115010

Zhang Y, Feng T, Wang L, Wang X, Wei T, Wang P (2023) Effects of long-term vegetation restoration on soil physicochemical properties mainly achieved by the coupling contributions of biological synusiae to the Loess Plateau. Ecol Indic 152:110353

Zhang Y, Balasubramanian D, Song Q, Zhou W (2024) Environmental and management controls of soil carbon storage in grasslands of southwestern China (No. EGU24-2782). Copernicus Meetings

Zhao B (1992) Effect of soil pH on the production of external hyphae by three VA mycorrhizal fungi. International Symposium on management of mycorrhizas in agriculture, horticulture, and forestry. Univ Western Aust Nedlands, W.A. p 36

Zhao WR, Shi RY, Hong ZN et al (2022) Critical values of soil solution Al activity and pH for canola and maize cultivation in two acidic soils. J Sci Food Agric 102(15):6984–6991

Zobeck TM, Bilbro JD (2001) Crop productivity and surface soil properties of a severely wind-eroded soil. In: Sustaining the global farm, selected papers from the 10th international soil conservation organization meeting. Int. Soil Conserv. Organisation, West Lafayette, pp 617–622

The manufacturer's authorised representative in the EU is Springer Nature Customer Service Centre GmbH, Europaplatz 3, 69115 Heidelberg, Germany. If you have any concerns regarding our products, please contact ProductSafety@springernature.com

Printed and bound by CPI Group (UK) Ltd, Croydon, CR0 4YY
30/09/2026
02225297-0002